3D Remote Sensing Applications in Forest Ecology

3D Remote Sensing Applications in Forest Ecology

Composition, Structure and Function

Special Issue Editors

Hooman Latifi
Ruben Valbuena

MDPI • Basel • Beijing • Wuhan • Barcelona • Belgrade

MDPI

Special Issue Editors

Hooman Latifi
Faculty of Geodesy and
Geomatics Engineering, K. N.
Toosi University of Technology
Iran

Ruben Valbuena
School of Natural Sciences,
Bangor University
UK

Editorial Office
MDPI
St. Alban-Anlage 66
4052 Basel, Switzerland

This is a reprint of articles from the Special Issue published online in the open access journal *Forests* (ISSN 1999-4907) from 2018 to 2019 (available at: https://www.mdpi.com/journal/forests/special_ issues/3D_Remote_Sensing).

For citation purposes, cite each article independently as indicated on the article page online and as indicated below:

LastName, A.A.; LastName, B.B.; LastName, C.C. Article Title. *Journal Name* **Year**, *Article Number*, Page Range.

ISBN 978-3-03921-782-3 (Pbk)
ISBN 978-3-03921-783-0 (PDF)

Cover image courtesy of Hooman Latifi.

Contents

About the Special Issue Editors

Hooman Latifi is a double-affiliated scholar of the K. N. Toosi University of Technology (Dept. of Photogrammetry and Remote Sensing) in Tehran-Iran and the University of Würzburg (Dept. of Remote Sensing) in Germany. He received his PhD from the Albert-Ludwigs-University Freiburg with a full scholarship funded by the German Academic Exchange Service (DAAD) in November 2011. During 2012 and 2018 he was a post-doc researcher at the Dept. of Remote Sensing of the Ludwigs-Maximilians-University Würzburg, established in cooperation with the German Aerospace Center (DLR). He received a degree known as "Privatdozent" (Associate Professor) in February 2018. In addition, he acted as the spokesman of the working group Ecology and Environment of the German region of the International Biometric Society during 2013 and 2017. He was a visiting scientist at the Dept. of Forestry of Michigan State University in the United States (2011) and the University of Angers in France (2014–2016). His research currently focuses on the applied spatial analysis of forest entities (structure, biodiversity, and health indicators) by means of spaceborne and airborne (mainly laser scanner and UAV) remote sensing data. In addition to being the author and co-author of several peer-reviewed publications in top-ranked journals of remote sensing, physical geography, forestry and environmental sciences, he has edited Special Issues of Forests (MDPI), Remote Sensing (MDPI) and PFG (Springer) journals on forestry applications of remote sensing. He is currently a member of the Editorial Board of Forestry (Oxford). Google Scholar profile at: https://scholar.google.de/citations?user=Y2CMsT0AAAAJ&hl=en.

Rubén Valbuena is a Lecturer in Forest Science at Bangor University (UK). Previously, he was a Research Fellow at the Department of Plant Sciences of the University of Cambridge, working under the Marie S. Curie Individual Fellowship project LORENZLIDAR. He holds the title of Docent (Adjunct Professor, 2017) in Remote Sensing and Biodiversity Indicators, and DSc (2015) in Agriculture and Forest Science from the School of Forest Sciences at University of Eastern Finland (UEF). He obtained a PhD (2013) in the field of Remote Sensing, an MSc (2007) in Environmental Science from the Forestry College at Technological University of Madrid (UPM), and a BSc (2002) in Biology from University of Navarra (Spain). He has participated in numerous projects involving forest ecology and remote sensing, among them, assessing the feasibility of LIDAR to retrieve essential biodiversity variables at the UN Environment-World Conservation Monitoring Centre (UNEP-WCMC) (2017, 2018), developing Pan-European indicators of forest structure at the European Forest Institute (EFI; 2010–2014), fuelwood mapping in Burkina Faso (UEF, 2014), priority habitat mapping in Wales, remote sensing-assisted forest inventory in Scotland at the Forest Research Agency of the UK Forestry Commission (2008, 2009), and obtaining baseline indicators for strategic environmental assessment at UPM (2006–2008).

Preface to "3D Remote Sensing Applications in Forest Ecology"

Dear Colleagues, The composition, structure and function of forest ecosystems are the key features characterizing their ecological properties, and can thus be crucially shaped and changed by various biotic and abiotic factors, ranging from global, continental and sub-continental climate change to macro- and micro-climatic regimes, disturbance agents and anthropogenic factors. The constant and alarming rise in the magnitude and extent of these changes in recent decades calls for enhanced cross-border and cross-continental mitigation and adaption measures, which will entail intensified monitoring in both space and time. In the absence or shortage of expensive logistics and field surveys, remote sensing data and methods are the main complementary sources of up-to-date synoptic and objective information for forest ecology. Owing to the fact that forest ecosystems (and the influential factors shaping them) are inherently of a three-dimensional nature, the methods based on the analysis of three-dimensional sources of remote sensing data can be considered the most appropriate tools to resemble the forest compositional, structural, and functional dynamics. Examples of these data embrace a broad range of methods for 3D reconstruction (stereo-photogrammetric restitution, structure from motion, interferometry, ranging, etc.) obtained using various remote sensors (digital images, LIDAR, or RADAR) from a variety of platforms (ground-based, UAV-borne, airborne, or spaceborne). While many applications rely on the sole use of either of these data sources to answer a specific question, combined or fused applications have received considerable attention in recent years. In this Special Issue of Forests, we published a set of state-of-the-art scientific works from a rather wide range of experimental studies, method developments and model validations, all dealing with the general topic of 3D remote sensing-assisted applications in monitoring forest composition, structure, and function. We aimed to demoonstrate applications in forest ecology from a broad collection of method/sensor/platform combinations, including fusion schemes. The published papers detail works from seven countries located on three continents and embracing multiple biomes and ecosystems. They also cover the application of multiple remote sensing and geospatial data sources, ranging from optical to synthetic aperture radar and laser scanner data. All in all, the studies and their focuses were as broad as the forest ecological applications of remote sensing itself and, thus, reflect the current and very diverse usages, as well as directions, in which future research and practical works will go.

Hooman Latifi, Ruben Valbuena
Special Issue Editors

Editorial

Current Trends in Forest Ecological Applications of Three-Dimensional Remote Sensing: Transition from Experimental to Operational Solutions?

Hooman Latifi [1,2,*] and **Ruben Valbuena** [3]

1   Faculty of Geodesy and Geomatics Engineering, K.N. Toosi University of Technology, P.O. Box 15875-4416, Tehran, Iran
2   Department of Remote Sensing, University of Würzburg, Oswald Külpe Weg 86, 97074 Würzburg, Germany
3   School of Natural Sciences, Bangor University, Bangor, Gwynedd LL57 2UW, UK; r.valbuena@bangor.ac.uk
*   Correspondence: hooman.latifi@kntu.ac.ir; Tel.: +98-(0)21-88877070-3 (ext. 312)

Received: 30 September 2019; Accepted: 8 October 2019; Published: 9 October 2019

check for updates

Abstract: The alarming increase in the magnitude and spatiotemporal patterns of changes in composition, structure and function of forest ecosystems during recent years calls for enhanced cross-border mitigation and adaption measures, which strongly entail intensified research to understand the underlying processes in the ecosystems as well as their dynamics. Remote sensing data and methods are nowadays the main complementary sources of synoptic, up-to-date and objective information to support field observations in forest ecology. In particular, analysis of three-dimensional (3D) remote sensing data is regarded as an appropriate complement, since they are hypothesized to resemble the 3D character of most forest attributes. Following their use in various small-scale forest structural analyses over the past two decades, these sources of data are now on their way to be integrated in novel applications in fields like citizen science, environmental impact assessment, forest fire analysis, and biodiversity assessment in remote areas. These and a number of other novel applications provide valuable material for the *Forests* special issue "3D Remote Sensing Applications in Forest Ecology: Composition, Structure and Function", which shows the promising future of these technologies and improves our understanding of the potentials and challenges of 3D remote sensing in practical forest ecology worldwide.

Keywords: 3D remote sensing; composition; forest ecology; function; structure

1. Introduction

The research on understanding the underlying ecological processes of forest ecosystems has been amongst the main interests in natural sciences for centuries. A high number of text books written by forest ecologists on forest ecology are available, in which basic concepts (e.g., ecological functions, interrelated patterns, flora, fauna and their dynamics) and detailed topics (e.g., connection to other ecological branches like community or population ecology, energy flux, complexity and regeneration patterns in forest ecosystems) are elaborated either as a whole [1–3] or by considering specific global biome- and ecosystem-specific characteristics [4–6]. However, one may note that common ecological concepts like biodiversity, ecosystem functioning and structure are overly multi-dimensional and cannot be subject to crisp definitions [7].

By defining a framework to answer most ecological questions one may, however, also note the structure of forest landscapes in general, which is inherently complex and three-dimensional (3D). This is mainly raised by the presence and dynamics of vegetative elements that harmonize with factors like topography, wildlife and climatic variables. This complexity has necessitated that researchers selectively focus on a subset of forest ecological components, while neglecting others in a given research framework [8]. Regardless of the degree in which problems concerning forest structural, compositional and functional traits are simplified, the 3D nature of forest ecosystems stands as one of the most essential aspects influencing almost the entire ecosystem dynamics, and should therefore be given the highest consideration.

Bearing this in mind and given the tremendous difficulties associated with the logistics, manpower and temporal repeatability of field-based surveys for forest ecological research, various available sources of data acquired by space-borne, air-borne and terrestrial remote sensing sensors have nowadays become indispensable sources of information for research on spatiotemporal dynamics of forest ecosystems. However, here we deliberately focused on 3D sources of data due to (1) their higher semantic association with most concepts and attributes in forest ecology and (2) the existing dearth of collective research summary (e.g., reviews, proceedings and journal special issues) on their applications in forest ecology.

There are currently several sources of remotely sensed 3D data available that can be useful for forest applications. Space-borne sources range from stereo pairs of optical, multi-angular, satellite sensors [9,10] to synthetic aperture radar (SAR)-based measurements [11,12] and space-borne laser scanning [13]. The airborne sources are much more diverse, including airborne laser scanning [14–16], airborne SAR [17], and traditional stereo airborne photogrammetry [18,19]. These also include many of those surveyed from unmanned aerial vehicles (UAVs), for which structure from motion [20] has become a predominant source of 3D information, while the availability of light-weight devices like LiDAR [21] and other nanosensors improves each year. There is also a large range of plausible combinations of sensors to measure or estimate forest variables from terrestrial platforms (terrestrial laser scanning [22,23], portable profiling lidar [24], fish-eye [25] and traditional stereo-photogrammetry [26], and even GPS receivers [27]). Each of them has distinct limitations in spatial and temporal coverages and the associated costs. Whereas each data source, individually or as categorized (optical stereoscopic, interferometric or polarimetric SAR and laser scanning), is associated with its specific analytical data processing, pros and cons, what they all have in common is that they reflect the 3D nature of forest ecosystems on different levels and are thereby currently of great interest for both forest science and practical forestry. Despite the rather long-studied and conventional application of 3D information in fields such as predictive modeling of forest structural attributes [14] and modeling biodiversity measures like abundance and occurrence of animal assemblages and habitat characterization [28], forest fire regimes [29], or environmental impact assessment of civil engineering projects, the use of 3D information in forest landscapes is still considered novel and many aspects require further investigation. In addition, attention has been recently drawn to integrating 3D data and methods in practical forest ecosystem survey and management, which is mainly motivated by (1) reduced costs of data acquisition (in particular terrestrial and airborne data) [30] and (2) integration and fusion of data from multiple sources, including spatially high-resolution 3D data and spatially-extensive, and often free-of-charge, multispectral optical data using novel algorithms [31].

2. Summary of the Contributions

As mentioned above, there is currently a lack of collective research reports, yet an increasing interest on novel and integrative researches on forest ecology by means of 3D sources of remote sensing data. Thus, the *Forests* special issue "3D Remote Sensing Applications in Forest Ecology: Composition, Structure and Function" was conceptualized by the authors of this paper and finally hosted 10 peer-reviewed contributions in which 3D sources of remote sensing data were applied either as a preliminary or auxiliary sources of information to understand, classify, augment, model and predict forest ecological attributes. Geographically, the contributions published within this special issue were well distributed around the globe, including China (four contributions) [32–35], Canada [36], Germany [37], India [38], Iran [39], Panama [40] and the United States [41]. The geographical distribution of the countries in which the published contributions were carried out are summarized in Figure 1.

In terms of global climatic regimes and ecological biomes, the temperate biome included the majority of works with seven studies [33–37,39,41], followed by sub-tropical [32,38] and tropical [40] biomes.

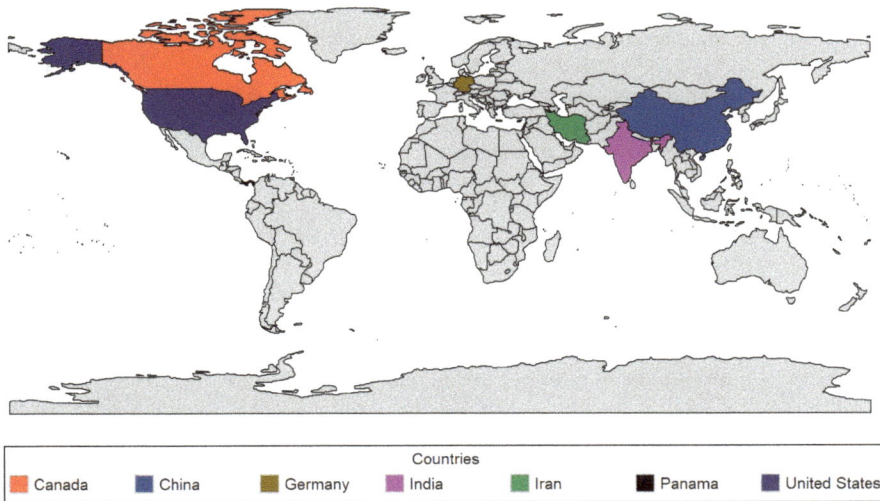

Figure 1. Distribution of the countries in which the study sites were located.

The topics covered within the published contributions can be divided into multiple groups: There were studies with rather classical applications such as single tree-level prediction of forest structural attributes by terrestrial laser scanning or visual estimation from Google Street View [33,41] and area-based prediction of forest structural attributes by space-borne stereo imagery, laser scanning or combination of passive optical with multi-frequency SAR data [34,35,39]. As an example, Ataee et al. [39] proved that a combination of space-borne SAR and optical data could improve performance and reduce uncertainties in the retrieval of tree volume. A number of works conducted on novel domains were also published, including a correlation between forest spectral burned ratios and height metrics derived from terrestrial laser scanning (using rather conventional height metrics for a novel application) [36], followed by other papers on hitherto rarely-studied topics like association between post-harvest tree root collar geometry and stump height by terrestrial laser scanning [37] and combining space-borne spectral and 3D data for fractional cover mapping of invasive alien woody species [38]. Moreover, Vallejos et al. [40] focused on a crucial, yet often neglected, source of statistical problem caused when working with optical remote sensing on quantitative ecological data, co-dispersion errors and data noise, whose results can be directly generalized to any existing source of 3D data. Similarly, habitat fragmentation caused by civil projects in forest ecosystems was surveyed by Li et al. [32], who addressed a generally remarkable topic that can be extended to similar cases using or combining 3D data sources like UAV-borne digital surface models. Here, the editors were open to those submissions with the main rationale that covering such crucial but still marginal topics might succeed in motivating extended research conducted on real 3D data. All in all, the published papers followed no biased tendency towards any specific group of relevant methodical or data-driven topics, but care was instead taken to host a collection of common applications that are currently in transition from being pure experimental to being implemented by the practitioners, together with those that are currently subject to no intensive research but contain great potential to be further addressed by the research community.

In terms of forest ecological attributes, the published papers represent a wide variety of attributes and thereby reflect the utterly diverse range of forest ecological attributes. As partly discussed above, the variables range from continuous variables that are commonly subject to regression modeling (allometric tree and stand structural attributes, root geometry, edaphic variables) [33–35,37,39–41] to categorical attributes that are relevant for classification approaches (fractional cover estimations of invasive species, landscape fragmentation) [32,38] and even to the use of vegetation indices and ratios [32,36]. One may, however, note that the current and potential applications of 3D remote sensing data in forest ecological domain go far beyond those covered here, with some examples being characterization of flora and fauna, coastal ecosystems, plant biodiversity (i.e., species richness), abiotic and biotic forest disturbances, sample size/sampling grid optimization for reference data surveys and many more. Therefore, we encourage further special issues focusing on publishing works on those subjects, in particular using state-of-the-art sources of 3D data such as space-borne laser scanning (e.g., ICESat-2 GLAS altimetry), space-borne C and L band interferometry (e.g., Sentinel and ALOS-2 data) and very high spatial resolution space-borne stereo optical data (e.g., SPOT 6-7, Pléiades 1A/1B, KOMPSAT series as well as SuperView-1 data from the GaoJing satellite).

A future pillar of research should also specifically concentrate on strengthening data assimilation and integrated use of multiple high- and medium-spatial resolution data sources. Recently published examples of such data assimilations are UAV with freely available optical data [42] and terrestrial laser scanning with multimodal space-borne data [43] for retrieving forest structural attributes. For forest ecological applications, this would concretely mean enhanced potentials for important practical applications like large-area calibrations of local models, monitoring remote and inaccessible mountainous forest ecosystems, calibrating small-area observations with large-area data on animal movements, and studying large-area habitat fragmentations. Therefore, we strongly encourage the remote sensing and forest ecological communities to intensify work transitioning from pure experimental data and methods to large-area practical applications, which could soon get enough popularity to be a topic for a future special issue of *Forests*.

Finally, this special issue was privileged by the high visibility and credibility of *Forests* in the Open Access domain to host a series of high-quality papers conducted by renowned international researchers. Nevertheless, the authors of this editorial treasure this opportunity to welcome future calls for similar relevant topics with a more practical orientation, possibly with a degree of financial incentives offered by MDPI to reduce publication fees. This would further support hosting quality research works and would enhance the visibility of both *Forests* and the published remote sensing/forest ecological works therein.

Acknowledgments: We thank all authors who contributed to this special issue by their papers, as well as all the reviewers who maintained the quality standard of this special issue of *Forests* by their timely, detailed and reliable reviews. The editorial team of MDPI, in particular Danae Yu, is also appreciated for their timely and flexible communication concerning the submission, review and publication process.

Conflicts of Interest: The authors declare no conflict of interest.

References

1. Kimmins, J.P. *Forest Ecology*, 3rd ed.; Benjamin Cummings: San Francisco, CA, USA, 2003; 720p, ISBN 9780130662583.
2. Peh, K.S.H.; Corlett, R.T.; Bergeron, Y. *Routledge Handbook of Forest Ecology*, 1st ed.; Routledge: London, UK, 2015; 656p, ISBN 9781315818290.
3. Van der Valk, A.G. *Forest Ecology: Recent Advances in Plant Ecology*; Springer: Cham, The Netherlands, 2009; 363p, ISBN 978-90-481-2794-8.
4. Montagnini, F.; Jordan, C.F. *Tropical Forest Ecology: The Basis for Conservation and Management*, 1st ed.; Springer: Berlin/Heidelberg, Germany, 2005; 295p, ISBN 978-3-540-23797-6.
5. Roda, F.; Ratena, J.; Gracia, C.A.; Bellot, J. *Ecology of Mediterranean Evergreen Oak Forests*; Springer: Berlin/Heidelberg, Germany, 1999; 377p, ISBN 978-3-642-63668-4.
6. Larsen, J.A.; Kozlowski, T.T. *The Boreal Ecosystem*, 1st ed.; Academic Press: Cambridge, MA, USA, 1980; 516p, ISBN 9781483269870.

7. Wang, K.; Franklin, S.E.; Guo, X.; Cattet, M. Remote Sensing of Ecology, Biodiversity and Conservation: A Review from the Perspective of Remote Sensing Specialists. *Sensors* **2010**, *10*, 9647–9667. [CrossRef] [PubMed]

8. Lowman, M.D.; Rinker, H.B. *Forest Canopies*, 2nd ed.; Academic Press: Cambridge, MA, USA, 2004; 544p, ISBN 978-0-12-457553-0.

9. Maack, J. Modeling forest biomass using Very-High-Resolution data—Combining textural, spectral and photogrammetric predictors derived from spaceborne stereo images. *Eur. J. Remote Sens.* **2015**, *48*, 245–261. [CrossRef]

10. Toutin, T.; Schmitt, C.; Wang, H. Impact of no GCP on elevation extraction from WorldView stereo data. *ISPRS J. Photogramm. Remote Sens.* **2012**, *72*, 73–79. [CrossRef]

11. Park, S.E. The Effect of Topography on Target Decomposition of Polarimetric SAR Data. *Remote Sens.* **2015**, *7*, 4997–5011. [CrossRef]

12. Yin, J.; Moon, W.M.; Yang, J. Novel Model-Based Method for Identification of Scattering Mechanisms in Polarimetric SAR Data. *IEEE Trans. Geosci. Remote Sens.* **2016**, *54*, 520–532. [CrossRef]

13. Silva, C.A.; Saatchi, S.; García, M.; Labriere, N.; Klauberg, C.; Ferraz, A.; Meyer, V.; Jeffery, K.J.; Abernethy, K.; White, L.; et al. Comparison of Small- and Large-Footprint Lidar Characterization of Tropical Forest Aboveground Structure and Biomass: A Case Study From Central Gabon. *IEEE J. Sel. Top. Appl. Earth Obs. Remote Sens.* **2018**, *11*, 3512–3526. [CrossRef]

14. Maltamo, M.; Mehtätalo, L.; Valbuena, R.; Vauhkonen, J.; Packalen, P. Airborne Laser Scanning for Tree Diameter Distribution Modelling: A Comparison of Modelling Alternatives in a Tropical Single-Species Plantation. *Forestry* **2018**, *91*, 121–131. [CrossRef]

15. Bottalico, F.; Chirici, G.; Giannini, R.; Mele, S.; Mura, M.; Puxeddu, M.; McRoberts, R.E.; Valbuena, R.; Travaglini, D. Modeling Mediterranean forest structure using airborne laser scanning data. *Int. J. Appl. Earth Obs. Geoinf.* **2017**, *57*, 145–153. [CrossRef]

16. Valbuena, R.; Maltamo, M.; Packalen, P. Classification of Multi-Layered Forest Development Classes from Low-Density National Airborne Lidar Datasets. *Forestry* **2016**, *89*, 392–401. [CrossRef]

17. Tanase, M.A.; Panciera, R.; Lowell, K.; Tian, S.; Hacker, J.M.; Walker, J.P. Airborne multi-temporal L-band polarimetric SAR data for biomass estimation in semi-arid forests. *Remote Sens. Environ.* **2014**, *145*, 93–104. [CrossRef]

18. Hinsken, L.; Miller, S.; Tempelmann, U.; Uebbing, R.; Walker, S. Triangulation of LH Systems ADS40 Imagery Using Orima GPS/IMU. *Int. Arch. Photogramm. Remote Sens. Spat. Inf. Sci.* **2002**, *34*, 156–162.

19. Hofmann, O.; Nave, P.; Ebner, H. DPS—A Digital Photogrammetric System for Producing Digital Elevation Models and Orthophotos by Means of Linear Array Scanner Omagery. *Photogramm. Eng. Remote Sens.* **1984**, *50*, 1135–1142.

20. Hentz, A.M.K.; Silva, C.A.; Dalla Corte, A.P.; Netto, S.P.; Strager, M.P.; Klauberg, C. Estimating Forest Uniformity in Eucalyptus spp. and Pinus taeda L. Stands Using Field Measurements and Structure from Motion Point Clouds Generated from Unmanned Aerial Vehicle (UAV) Data Collection. *For. Syst.* **2018**, *27*, e005. [CrossRef]

21. Almeida, D.; Broadbent, E.; Zambrano, A.; Wilkinson, B.; Ferreira, M.; Chazdon, R.; Meli, P.; Gorgens, E.; Silva, C.; Stark, S.; et al. Monitoring the structure of forest restoration plantations with a drone-lidar system. *Int. J. Appl. Earth Obs. Geoinf.* **2019**, *79*, 192–198. [CrossRef]

22. Henning, J.G.; Radtke, P.J. Detailed Stem Measurements of Standing Trees from Ground-Based Scanning Lidar. *For. Sci.* **2006**, *52*, 67–80.

23. Liang, X.; Kankare, V.; Hyyppä, J.; Wang, Y.; Kukko, A.; Haggrén, H.; Yu, X.; Kaartinen, H.; Jaakkola, A.; Guan, F.; et al. Terrestrial laser scanning in forest inventories. *ISPRS J. Photogramm. Remote Sens.* **2016**, *115*, 63–77. [CrossRef]

24. Almeida, D.R.A.; Nelson, B.W.; Schietti, J.; Görgens, E.B.; Resende, A.F.; Starkm, S.C.; Valbuena, R. Contrasting Fire Susceptibility and Fire Damage Between Seasonally Flooded Forest and Upland Forest in the Central Amazon Using Portable Terrestrial Profiling Lidar. *Remote Sens. Environ.* **2016**, *184*, 153–160. [CrossRef]

25. Herrera, P.J.; Pajares, G.; Guijarro, M.; Ruz, J.J.; Cruz, J.M.; Montes, F. A Featured-Based Strategy for Stereovision Matching in Sensors with Fish-Eye Lenses for Forest Environments. *Sensors* **2009**, *9*, 9468–9492. [CrossRef]

26. Forsman, M.; Börlin, N.; Holmgren, J. Estimation of tree stem attributes using terrestrial photogrammetry. *ISPRS Int. Arch. Photogramm. Remote Sens. Spat. Inf. Sci.* **2012**, 261–265. [CrossRef]

27. Mollfulleda, A.; Martin, F.; Paloscia, S.; Santi, E.; Guerriero, L.; Pierdicca, N.; Floury, N. GNSSBio: Forest Biomass Retrieval Based on GNSS Ground Receiver. In Proceedings of the 2017 IEEE International Geoscience and Remote Sensing Symposium (IGARSS), Institute of Electrical and Electronics Engineers (IEEE), Fort Worth, TX, USA, 23–28 July 2017; pp. 5778–5781.

28. Mononen, L.; Auvinen, A.P.; Packalen, P.; Virkkala, R.; Valbuena, R.; Bohlin, I.; Valkama, J.; Vihervaara, P. Usability of citizen science observations together with airborne laser scanning data in determining the habitat preferences of forest birds. *For. Ecol. Manag.* **2018**, *430*, 498–508. [CrossRef]

29. McCarley, T.R.; Kolden, C.A.; Vaillant, N.M.; Hudak, A.T.; Smith, A.M.; Wing, B.M.; Kellogg, B.S.; Kreitler, J. Multi-temporal LiDAR and Landsat quantification of fire-induced changes to forest structure. *Remote Sens. Environ.* **2017**, *191*, 419–432. [CrossRef]

30. Clementel, F.; Colle, G.; Farruggia, C.; Floris, A.; Scrinzi, G.; Torresan, C. Estimating forest timber volume by means of "low-cost" LiDAR data. *Ital. J. Remote Sens.* **2012**, *44*, 125–140. [CrossRef]

31. Valbuena, R.; Mauro, F.; Arjonilla, F.J.; Manzanera, J.A. Comparing airborne laser scanning-imagery fusion methods based on geometric accuracy in forested areas. *Remote Sens. Environ.* **2011**, *115*, 1942–1954. [CrossRef]

32. Li, X.; Lin, Y. Do High-Voltage Power Transmission Lines Affect Forest Landscape a d Vegetation Growth: Evidence from a Case for Southeastern of China. *Forests* **2019**, *10*, 162. [CrossRef]

33. Liu, G.; Wang, J.; Dong, P.; Chen, Y.; Liu, Z. Estimating Individual Tree Height and Diameter at Breast Height (DBH) from Terrestrial Laser Scanning (TLS) Data at Plot Level. *Forests* **2018**, *9*, 398. [CrossRef]

34. Liu, M.; Cao, C.; Dang, Y.; Ni, X. Mapping Forest Canopy Height in Mountainous Areas Using ZiYuan-3 Stereo Images and Landsat Data. *Forests* **2019**, *10*, 105. [CrossRef]

35. Zhang, Y.; Shi, Y.; Choi, S.; Ni, X.; Myneni, R.B. Mapping Maximum Tree Height of the Great Khingan Mountain, Inner Mongolia Using the Allometric Scaling and Resource Limitations Model. *Forests* **2019**, *10*, 380. [CrossRef]

36. Kato, A.; Moskal, L.M.; Batchelor, J.L.; Thau, D.; Hudak, A.T. Relationships between Satellite-Based Spectral Burned Ratios and Terrestrial Laser Scanning. *Forests* **2019**, *10*, 444. [CrossRef]

37. Labelle, E.R.; Heppelmann, J.B.; Borchert, H. Application of Terrestrial Laser Scanner to Evaluate the Influence of Root Collar Geometry on Stump Height after Mechanized Forest Operations. *Forests* **2018**, *9*, 709. [CrossRef]

38. Khare, S.; Latifi, H.; Rossi, S.; Ghosh, S.K. Fractional Cover Mapping of Invasive Plant Species by Combining Very High-Resolution Stereo and Multi-Sensor Multispectral Imageries. *Forests* **2019**, *10*, 540. [CrossRef]

39. Ataee, M.S.; Maghsoudi, Y.; Latifi, H.; Fadaie, F. Improving Estimation Accuracy of Growing Stock by Multi-Frequency SAR and Multi-Spectral Data over Iran's Heterogeneously-Structured Broadleaf Hyrcanian Forests. *Forests* **2019**, *10*, 641. [CrossRef]

40. Vallejos, R.; Buckley, H.; Case, B.; Acosta, J.; Ellison, A.M. Sensitivity of Codispersion to Noise and Error in Ecological and Environmental Data. *Forests* **2018**, *9*, 679. [CrossRef]

41. Berland, A.; Roman, L.A.; Vogt, J. Can Field Crews Telecommute? Varied Data Quality from Citizen Science Tree Inventories Conducted Using Street-Level Imagery. *Forests* **2019**, *10*, 349. [CrossRef]

42. Puliti, S.; Saarela, S.; Gobakken, T.; Næsset, E. Combining UAV and Sentinel-2 auxiliary data for forest growing stock volume estimation through hierarchical model-based inference. *Remote Sens. Environ.* **2018**, *204*, 485–497. [CrossRef]

43. Mulatu, K.A.; Decuyper, M.; Brede, B.; Kooistra, L.; Reiche, J.; Mora, B.; Herold, M. Linking Terrestrial LiDAR Scanner and Conventional Forest Structure Measurements with Multi-Modal Satellite Data. *Forests* **2019**, *10*, 291. [CrossRef]

![forests logo]

forests

MDPI

Article

Improving Estimation Accuracy of Growing Stock by Multi-Frequency SAR and Multi-Spectral Data over Iran's Heterogeneously-Structured Broadleaf Hyrcanian Forests

Mohammad Sadegh Ataee [1], Yasser Maghsoudi [1,]* [ORCID], Hooman Latifi [1,2,]* [ORCID] and Farhad Fadaie [3]

[1] Department of Photogrammetry and Remote Sensing, Faculty of Geodesy and Geomatics Engineering, K, N. Toosi University of Technology, P.O.Box 15433-19967 Tehran, Iran

[2] Department of Remote Sensing, University of Würzburg, Oswald KülpeWeg 86, 97074 Würzburg, Germany

[3] Department of Forestry, University of Guilan, Entezam Square, P.O.Box 43619-96196 Some'e Sara, Iran

* Correspondence: ymaghsoudi@kntu.ac.ir (Y.M.); hooman.latifi@kntu.ac.ir (H.L.);
Tel.: +98-21-8887070-3 (ext. 223) (Y.M.); +98-21-8887070-3 (ext. 312) (H.L.)

Received: 26 June 2019; Accepted: 25 July 2019; Published: 29 July 2019

check for updates

Abstract: Via providing various ecosystem services, the old-growth Hyrcanian forests play a crucial role in the environment and anthropogenic aspects of Iran and beyond. The amount of growing stock volume (GSV) is a forest biophysical parameter with great importance in issues like economy, environmental protection, and adaptation to climate change. Thus, accurate and unbiased estimation of GSV is also crucial to be pursued across the Hyrcanian. Our goal was to investigate the potential of ALOS-2 and Sentinel-1's polarimetric features in combination with Sentinel-2 multi-spectral features for the GSV estimation in a portion of heterogeneously-structured and mountainous Hyrcanian forests. We used five different kernels by the support vector regression (nu-SVR) for the GSV estimation. Because each kernel differently models the parameters, we separately selected features for each kernel by a binary genetic algorithm (GA). We simultaneously optimized R^2 and RMSE in a suggested GA fitness function. We calculated R^2, RMSE to evaluate the models. We additionally calculated the standard deviation of validation metrics to estimate the model's stability. Also for models over-fitting or under-fitting analysis, we used mean difference (MD) index. The results suggested the use of polynomial kernel as the final model. Despite multiple methodical challenges raised from the composition and structure of the study site, we conclude that the combined use of polarimetric features (both dual and full) with spectral bands and indices can improve the GSV estimation over mixed broadleaf forests. This was partially supported by the use of proposed evaluation criterion within the GA, which helped to avoid the curse of dimensionality for the applied SVR and lowest over estimation or under estimation.

Keywords: GSV; nu SVR; uneven-aged mountainous; polarimetery; multi-spectral; optimization

1. Introduction

Hyrcanian forests are known as remnants of the Pleistocene era that survived the frost period [1]. These forests are located in regions of northern Iran and part of Caucasus, and embrace a high species and structural diversity of uneven-aged mountainous broadleaf forests distributed across a high altitudinal gradient [1,2]. Recently, portions of these forests were inscribed in the list of UNESCO World Natural Heritages [3]. The growing stock volume (GSV) is one of the important allometric biophysical forest attributes. It is closely related to other forest quantities such as height and aboveground biomass and is of great importance in the forest ecology, management, and carbon storage [4,5]. Tree-level biomass is conventionally derived by using species-specific allometric relations and wood density from ground-based measurements. However, the high cost, time, and the limited geographical coverage prohibitively challenge these methods. On the other hand, remote sensing data from spaceborne SAR and multispectral sensors with proper radiometric and spatial resolution and sufficient time intervals of data acquisition from the desired areas have been proven to provide important proxies in forestry research [6,7]. Due to mentioned historical and environmental reasons, development and implementation of remote sensing-assisted methods serve the overarching aim of monitoring and sustainable management of Hyrcanian forests.

Amongst the recent attempts for GSV estimation by state-of-the-art multispectral data, Chrysafis et al. (2017) [8] estimated the GSV by blending the Sentinel-2 and the Landsat data with the Random Forest (RF) model, and concluded that near-infrared and the red edge domains greatly affect GSV estimation. In addition, Mura et al. (2018) [9] estimated the GSV by the Sentinel-2, Landsat, and the Rapideye sensors and concluded that, beside the near-infrared and red edge regions, the SWIR region is also effective because of its sensitivity to the water content in the canopy. In both studies Sentinel-2 data were suggested to excess others in performance.

The radio detection and ranging (radar) sensors considerably contributed to solving the limitations of optical sensors including their inability to penetrate the canopy and less sensitivity to the vertical canopy structure. The ability of radar data to estimate biophysical forest characteristics is also less susceptible to weather conditions and acquisition time, which eases monitoring of mountainous forests that mostly occur in humid, cloudy, and foggy areas. However, changes in radar wavelength and type of polarization results in differences in both analytical workflow and the achieved estimation performance. Moreover, higher trunk volume leads to underestimation of actual GSV values due to the saturation in the scattering form dense canopy, which improves with an increase in wavelength [10–12]. Gao et al. (2018) [13] estimated GSV by the Dual polarized ALOS-1 data and reported a higher potential and later saturation of L-band HV cross-polarization channel than HH co-polarized channel. By using multi-temporal dual polarized ALOS-1 sensor data, Antropov et al. (2013) [14] concluded that the multi-temporal method was superior in prediction (with HH co-polarized channel performing better for mature trees), yet the saturation happens in the high GSV values. In addition, Chowdhury et al. (2014) [15] estimated the GSV by the ALOS-1 multi-temporal full polarimetric data, from which covariance and coherency matrices, as well as the phase difference between HH & VV channel and the coherency between HH and VV channels, were extracted. They showed that full polarimetry data has a high ability for GSV estimation.

Multi-sensor remote sensing approaches are highly capable for forest applications. In our study, we percept GSV estimation over mountainous broadleaf forests from a slightly different perspective. The biophysical characteristics of forest can be studied in multi-spectral approach by focusing on the biochemical aspects such as chlorophyll and in SAR approach on radar wave penetration in the canopy [16,17]. Mauya et al. (2019) [18] estimated GSV by the ALOS-2's global mosaic, the Sentinel-1 and the Sentinel-2 sensors data. They concluded that using SAR data alone was unlikely to provide a good estimation ability for GSV, while a combined use of Sentinel-1 and Sentinel-2 data were advantageous. In Iran's Hyrcanian region, recent investigations of the ability of remotely sensed data and methods include Vafaei et al. (2018) [19] who estimated biomass by the ALOS-2 full polarimetric and Sentinel-2A data. The Sentinel-2 returned the moderate accuracy, whereas the ALOS-2 individually led to minimum estimation accuracy.

The overall objective of this research was to estimate the GSV in Hyrcanian uneven-aged mountainous broadleaf forests based on leveraging a broad range of possibilities in optical and radar data processing. To this aim, we used ALOS-2 full polarimetric, Sentinel-1 dual polarimetric and Sentinel-2 multi-spectral data. We only concentrated on combined use of polarimetric and spectral features. In addition, we compared five different kernels in support vector regression (SVR) for GSV estimation. We additionally applied a heuristic feature selection by binary genetic algorithm, in which we simultaneously optimized the root mean square error (RMSE) and coefficient of determination

(R^2) for each kernel separately. The workflow and findings of this study are mainly significant due to multiple challenges associated with the structure and composition of our test site, including the severe topography, limited field samples, highly mixed tree species in various ages, and complex slope-aspect structure.

2. Materials and Methods

2.1. Study Area

The study area is located in Guilan province's Nav-Asalem region in northern Iran (see Figure 1). The altitude varies between 100 to 2000 m above sea level, and the slope ranges from 0 to 73 degrees. The climate is temperate and cold, average annual rainfall is 1200 mm per year and the mean temperature is 12.4-degrees Celsius. The forest comprises uneven-aged mountainous broadleaf stands dominated by oriental Beech (*Fagus orientalis* Lipsky.) and hornbeam (*Carpinus Betulus* L.), accompanied by other broadleaf shrub and tree species.

2.2. Field Data Inventory

Field data were collected during July 2014 to March 2015. The in situ measurements were conducted in 148 circular plots with 18 m radius each and distributed in a randomly positioned square grid, as prescribed by the technical bureau of the Iranian Forests, Rangelands and Watershed Management Organization (FRWO) [20]. In each plot, the diameter at breast height (DBH; i.e., 1.5 m above ground surface) and the species type were recorded for all trees. For a sample of trees, the dominant species was recorded in each plot. Finally, the GSV values were calculated in each stand by means of DBH and lookup tables for each species separately.

Plot center locations were collected by global positioning system (GPS) in the WGS84 coordinate system. The original values of GSV in each plot were multiplied by 10 in order to be transformed from m^3 per 0.1 hectares to m^3 per hectare. The inventoried GSV ranged between 98.60 and 385.8 m^3 ha^{-1} (see Table 1).

Table 1. Summary of descriptive statistics of GSV.

Descriptor	Value
Mean(m^3 ha^{-1})	247.86
Minimum(m^3 ha^{-1})	98.60
Maximum(m^3 ha^{-1})	385.8
Standard deviation(m^3 ha^{-1})	54.63
Number of plots	148

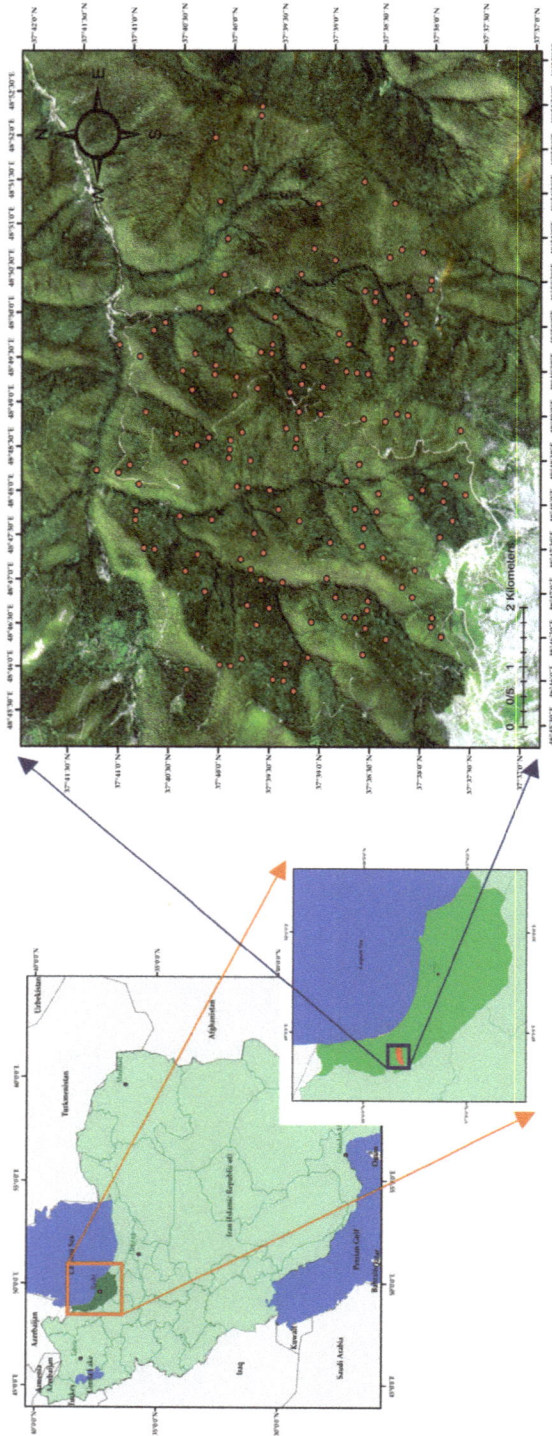

Figure 1. The geographical location of the study area in the WGS 84 coordinate system. The red-points display the field plots locations. Background image is Sentinel-2 RGB image (Red: Band 4; Green: band 3; Blue: band 2).

2.3. Remotely Sensed Data

2.3.1. Sentinel-2 Data

Sentinel-2 satellite carries a multi-spectral sensor in 13 bands from 400 nm to 2400 nm with spatial resolutions of 10, 20, and 60 m [21]. The 60-m resolution bands include B1 (430 nm), B9 (940 nm) and B10 (1340 nm) that are mainly used for atmospheric correction. The Sentinel-2 data in Level1C was acquired on 21 June 2016 and downloaded from European Space Agency (ESA) repository. After performing the atmospheric and geometric correction incorporating SRTM 1 aSec DEM [22], the spectral features were extracted from Sentinel-2 data. Since the acquired Level1C data represents the top-of-atmosphere (TOA) reflectance, we used Sen2Cor algorithm for atmospheric correction to calculate Level 2A data representing the bottom-of-atmosphere (BOA) reflectance. The sen2cor algorithm is an image-based correction that performs image correction based on lookup tables extracted from RadTran algorithm [23], with the main benefit that it does not need local meteorological data for correction. We only used the original bands and vegetation indices (see Table 2) for GSV estimation. Due to the different spatial resolutions of the Sentinel-2 imagery, we used the 5×5 local window for bands and indices with 10 m spatial resolution, whereas 3×3 local window was used for bands and indices with 20 m spatial resolution. The mean values of data extracted in each local window over each field plot inventory were applied for GSV estimation. The entire process was performed in SNAP V6 software [24].

Table 2. Sentinel-2 spectral features.

Feature	Used Bands	Resolution	Feature	Used Bands	Resolution
B2, B3, B4, B8	Original band	10 m	GEMI [25]	B8, B4	10 m
SAVI [26]	B8, B4	10 m	ARVI [27]	B2, B4, B8	10 m
TSAVI [28]	B8, B4	10 m	NDVI [8]	B4, B8	10 m
MSAVI [29]	B8, B4	10 m	B5, B6, B7, B8a	Original bands	20 m
			B11, B12		
MSAVI2 [29]	B8, B4	10 m	NDI45 [30]	B4, B5	20 m
DVI [8]	B8, B4	10 m	MTCI [31]	B4, B5, B6	20 m
RVI [32]	B8, B4	10 m	MCARI [33]	B3, B4, B5	20 m
PVI [8]	B8, B4	10 m	REIP [34]	B4, B5, B6, B7	20 m
IPVI [35]	B8, B4	10 m	S2REP [34]	B4, B5, B6, B7	20 m
WDVI [36]	B8, B4	10 m	IRECI [34,37]	B4, B5, B6, B7	20 m
TNDVI [32]	B8, B4	10 m	PSSRa [38]	B4, B7	20 m
GNDVI [39]	B3, B8	10 m			

B2 (490), B3 (560), B4 (665), B5 (705), B6 (740), B7 (783), B8 (842), B8a (865), B11 (1610), B12 (2100), Unit = nm.

2.3.2. Sentinel-1 Data

Sentinel-1 is equipped with a synthetic aperture radar (SAR) antenna that scans the Earth in dual polarization mode in VV and VH channels in the C band. Our study area was scanned in dual polarization mode on 22nd of July 2017 and data was downloaded from the ESA repository. The data was scanned in single look complex (SLC) mode using the interferometric wide (IW) method. The dataset was primarily calibrated. Since the data was in dual polarimetry mode, the calibration was performed in the complex calibration manner to preserve the phase information. For extracting polarimetric information, we first extracted the C2 matrix (non-coherent covariance matrix) from the calibrated data. For dual polarimetric data, we used the Equation (1) to calculate the C2 matrix [40]. Data was scanned by IW method and was thus required to be debursted by the Sentinel-1 TOPS deburst method. For squared pixels, we applied the multi-looking operator with five looks in the azimuth direction and one look in the range direction to reach the 17-meter resolution, i.e., the spatial resolution close to the field sample plot size. For protecting polarimetric information, we applied the polarimetric refined Lee filter (7×7 local window) on the deburst C2 matrix to decrease the speckle noise [39]. The range-doppler terrain correction method with the SRTM 1Sec HGT DEM [22] was used to georeference the C2 matrix. Finally, the dual-polarimetric H-A-Alpha decomposition was applied on the georeferenced C2 matrix [40]. Consequently, we used a 3 × 3 local window due to the available 17 m spatial resolution to extract information on sample plot level. The entire analysis was performed in SNAP V6 software [24]. The extracted features are summarized in Table 3.

$$C_2 = \left[\begin{array}{cc} \langle |S_{VV}|^2 \rangle & \langle S_{VV} S_{VH}^* \rangle \\ \langle S_{VH} S_{VV}^* \rangle & \langle |S_{VH}|^2 \rangle \end{array} \right] \tag{1}$$

Table 3. Sentinel-1 features.

Feature	Elements
C2 matrix	C11, C12_real, C12_image, C22
H/A/Alpha	Entropy, Anisotropy, Alpha

2.3.3. ALOS-2 PalSar Data

The ALOS-2 satellite features a fully polarimetric synthetic aperture radar (SAR) antenna. This sensor scans the Earth in HH, HV, VH, and VV polarization channel in L band. We downloaded the data acquired on 10th of June 2015 over our study site from the repository provided by the Japan Aerospace Exploration Agency (JAXA). This data was scanned in SLC mode with Stripmap-2 method. The data preprocessing was performed solely for the polarimetric features. For preserving phase information, the raw data was initially calibrated in the complex calibration manner. The T3 coherency matrix was extracted from the calibrated data (Equation (2)) [41]. One may note the negative effects caused by the severe topography on scattering from the scatterers and consequently on the polarimetric information [42]. One crucial effect of topography is the rotation of radar wave in the line of sight because of the slope in the azimuth direction. This phenomenon is known as polarization orientation angle (POA) and is compensated by applying Equations (3)–(5) [41–43]. The multi-looking operator with six looks in azimuth direction and four looks in range direction was applied to the T3 matrix for the squared pixel and speckle noise reduction. Finally, we reached the 15-m spatial resolution that was close to those of field sample plots. As already mentioned for Sentinel-1 data processing, we used refined Lee speckle filter to simultaneously reduce speckle noise and preserve polarimetric information [41,44,45]. The filter was applied with 7 × 7 local window, 3 × 3 target window, and sigma of 0.9 to the T3 matrix. Following these steps, the T3 matrix was consequently georeferenced by SRTM 1Sec HGT DEM [22] using the range-doppler terrain correction method. In this research, we only extracted the polarimetric features (see Table 4). We used a 3 × 3 local window to extract the information on sample plot level. Similar to the preceding steps, the processes were performed in SNAP V6 [24].

Table 4. ALOS-2 polarimeteric features.

Feature	Elements
H/A/Alpha [41]	Anisotropy, Alpha, Entropy, Beta, Delta, Gamma, Lambanda Alpha (1,2,3), Lambanda (1,2,3)
Yamaguchi [46]	Surface, Double, Volume, Helix
Van zyl [47]	Surface, Double, Volume
Cloude [41]	Surface, Double, Volume
Generalized Freeman-Durden [48]	Surface, Double, Volume
Touzi [49]	Psi, Tau, Phi, Alpha Psi (1,2,3),Tau (1,2,3), Phi (1,2,3), Alpha (1,2,3)
RVI [41]	RVI
SPAN [41]	SPAN
Pedestal height [41]	Pedestal height

$$S = \begin{bmatrix} S_{HH} & S_{HV} \\ S_{VH} & S_{VV} \end{bmatrix} . \vec{k} = \frac{1}{\sqrt{2}} \begin{bmatrix} S_{HH} + S_{VV} \\ S_{HH} - S_{VV} \\ 2S_{HV} \end{bmatrix} . T_3 = \vec{k}\,\vec{k}^{*T} \qquad (2)$$

$$\delta = \frac{1}{4} \left[\tan^{-1} \left(\frac{-4Re(\langle (S_{HH} - S_{VV})S_{HV}^* \rangle)}{-\langle |S_{HH} - S_{VV}|^2 \rangle + 4\langle |S_{HV}|^2 \rangle} \right) + \pi \right] \qquad (3)$$

$$V = \frac{1}{2} \begin{bmatrix} 1 + \cos 2\delta & \sqrt{2}\sin 2\delta & 1 - \cos 2\delta \\ -\sqrt{2}\sin 2\delta & 2\cos 2\delta & \sqrt{2}\sin 2\delta \\ 1 - \cos 2\delta & -\sqrt{2}\sin 2\delta & 1 + \cos 2\delta \end{bmatrix} \qquad (4)$$

$$T_3^{POA} = VT_3V^T \qquad (5)$$

In the equations S is the stocke's matrix, S_{HH} to S_{VV} are the four polarizations, k is the Pauli target vector for calculating T_3 matrix, and δ is the POA angle.

2.4. Modeling by Machine Learning

2.4.1. Support Vector Regression

Support vector machine (SVM) is known as powerful, flexible, and robust-to-noise machine learning method. The main goal of SVM is to minimize the sum of squares error (SSE) in Equation (6) during the model training procedure. The support vector regression (SVR) is a regression version of SVM [50,51]. There are two main methods for SVR, ε and υ regression. In ε regression, an insensitive tube defines by ε parameter, but this parameter is user-defined and the procedure affects model accuracy by only involving noisy observation in the big values of ε or the non-important observation data in the small values of ε, which results in non-generalized model that cannot model different behaviors of the phenomena [50,52,53]. In the υ regression method, the υ parameter defines the fraction of support vectors number involved in the modeling procedure [52]. In this method, the ε is automatically calculated in the algorithm and better controls support vectors. Therefore, both caveats of the ε method including challenge of the selection of proper ε value and its effect on accuracy could be resolved [52]. Therefore, we used the υ regression. The kernel based solution was used for accommodating the non-linear and complex behaviors. The used kernel passes the features into a new feature space [53], following which the model is trained in the new feature problem [51]. Whereas there are various kernels for SVR, the choice proper kernel for each problem depends on the nature of the problem [49]. We tested five different kernels (see Table 5). We tuned the hyper-parameters by repeated cross-validation based on RMSE (see Section 2.4.3). Besides the kernel's hyper-parameter, the cost parameter was also tuned with each kernel to maintain the model's flexibility [51].

$$\min \left(\frac{1}{2} \|w\|^2 + C \left(\upsilon \varepsilon + \frac{1}{l} \sum_{i=1}^{l} (\xi_i + \xi_i^*) \right) \right) \qquad (6)$$

In Equation (6), C regularization constant, υ parameter between 0 and 1, ε value and ξ_i^* is slack variable. We analyzed the results of three different experiments. In the first experiment, the GSV was modeled using a single sensor approach. In the second experiment, we modeled the GSV using all the features obtained through the multi-sensor approach. Finally, in the third experiment, we model the GSV using the GA selected features.

Table 5. Applied SVR kernels [54].

Kernel	Hyper-Parameter
Laplace	Sigma, Cost
RBF	Sigma, Cost
Polynomial	Degree, Offset, Scale, Cost
Sigmoid	Scale, Offset, Cost
Bessel	Sigma, Order, Degree, Cost

2.4.2. Feature Selection

Feature selection is an essential task in working with hyper-dimensional data, in which the best subset from all features is selected based on lower cost of model training and higher accuracy. This solution is essential in the presence of a large number of extracted features and small sample size to avoid the prohibitive effects of the curse of dimensionality on model performance and interpretability [55]. Therefore, a proper method for feature selection cares for the low training cost while avoiding the curse of dimensionality in the modeling [56]. The genetic algorithm (GA) is an evolutionary method for optimization. The method is inspired by natural selection in the real world, with the main idea of incorporating a wide range of suitable solutions that leads to selection of an optimum solution [57]. We used binary GA for feature selection. First, a population of chromosomes with 0 and 1 genes was randomly generated in which each chromosome is considered as a solution. The goodness of each solution was determined by the fitness function. The solutions with higher goodness had a higher chance to create a new solution. The new solution was then created by crossover procedure from two randomly selected solution with a high fitness value. The mutation operator changed the random genes with little probability and was used to maintain genetic diversity from one generation of a population of genetic algorithm chromosomes to the next. Finally, the algorithm was iterated until the fitness value remained unchanged [58]. Here, we selected the features for each kernel individually by cross-validation (see Section 2.4.3). The fitness function pursued the goal to simultaneously RMSE and R^2 values (see Figure 2). For this purpose, the fitness function as defined in Equations (7)–(9) was proposed. In this fitness function, the highest fitness value could potentially occur in R^2 equal to 1 and RMSE equal to 0, i.e., a fitness value equal to 190.

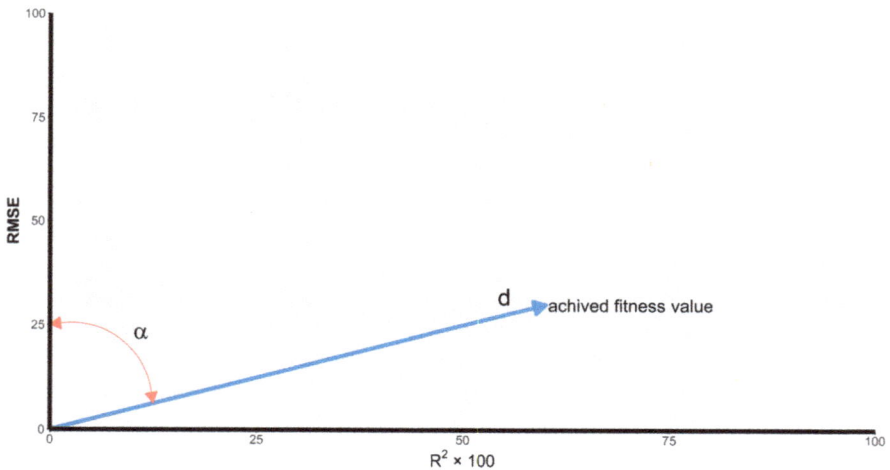

Figure 2. Applied fitness function.

$$\alpha = 90 - \tan^{-1}\left(\frac{RMSE}{100 \times R^2}\right) \quad (7)$$

$$d = \sqrt{RMSE^2 + (100 \times R^2)^2} \tag{8}$$

$$fitness = \alpha + d \tag{9}$$

2.4.3. Validation

The validation of model performance is the essential step in modeling [20]. Due to the most common case of limited sample size, using the k-fold cross-validation was necessary to avoid noisy and unstable predictions [18,50]. The RMSE, R^2 (Equations (10)–(11)) were used as error diagnostics based on their proven usefulness in assessing prediction of forest parameters [18,59]. In addition, the model stability in the presence of limited field sample size was checked by the repeated k-fold cross-validation. The model was assessed as more stable if the standard deviation (Equation (13)) of each of the diagnostic metrics was closer to zero [60]. Also we used MD (see Equation (12) for models over-estimation or under-estimation analysis. The higher positive values indicates the over-fitting and higher negative values indicates the under-fitting problem [61]. We used the 5-fold cross validation with 20 repetitions for more stable feature selection in the GA (Section 2.4.2), kernel's hyper-parameter tuning (Section 2.4.1) and finally, the model validation. The statistical modeling and the feature selections were implemented in open source domain in R using the kernlab, caret, and GA libraries.

$$RMSE = \sqrt{\sum_{i=1}^{n} \left(y_i^{obs} - y_i^{pred} \right)^2 / n} \tag{10}$$

$$R^2 = \sqrt{\sum_{i=1}^{n} \left(y_i^{pred} - \overline{y} \right)^2 / \sum_{i=1}^{n} \left(y_i^{obs} - \overline{y} \right)^2} \tag{11}$$

$$MD = \frac{1}{n} \sum_{i=1}^{n} \left(y_i^{pred} - y_i^{obs} \right) \tag{12}$$

$$SD = \sqrt{\sum_{i=1}^{m} \left(P_i - \overline{P} \right)^2 / m} \tag{13}$$

In the Equations, the y_i^{obs} is the observed GSV in plot i, y_i^{pred} is the predicted GSV, n is the number of samples in fold, $\overline{y}$ is the GSV's average in fold, P_i is the validation metric in each fold, $\overline{P}$ is the metric's average and m is the number by which the model was trained with different data.

3. Results

3.1. Single Sensor Models

3.1.1. Sentinel-2

Here, the 31 features extracted from the Sentinel-2 data were used for the GSV modeling (see Table 2). As previously mentioned in Section 2.3, the SVR model was employed using five different Kernels. The kernel's hyper-parameters were tuned with repeated cross-validation. The results can be seen in Table 6.

Table 6. Sentinel-2 modeling with 31 features.

Kernel	RMSE	R^2	MD	RMSE_SD	R^2_SD
Laplace	31.793	0.673	−0.003	2.919	0.071
RBF	31.644	0.673	−1.194	2.639	0.072
Sigmoid	31.693	0.674	−2.07	2.642	0.074
Polynomial	31.553	0.677	−0.62	2.630	0.071
Bessel	31.536	0.676	−1.38	2.662	0.072

As can be seen in Table 6, from the perspective of performance, the Bessel kernel with RMSE = 31.536 and R^2 = 0.676 provided a better performance compared to other kernels. The polynomial kernel presented a very close performance to the Bessel Kernel. From the perspective of model stability (see Section 2.3.3), the Polynomial kernel with RMSE_SD = 2.630 and R^2_SD = 0.071, provided more stable results than other kernels. The RBF kernel was the second more stable kernel (Table 6). The MD index shows that the all models are under estimate the GSV.

By considering both the models performance, stability and MD analysis the polynomial kernel exceeded other kernels, while all models generally showed close performance. Finally, no significant difference was observed amongst the kernels in the GSV modeling using the features extracted from Sentinel-2 (see Figures 3–5).

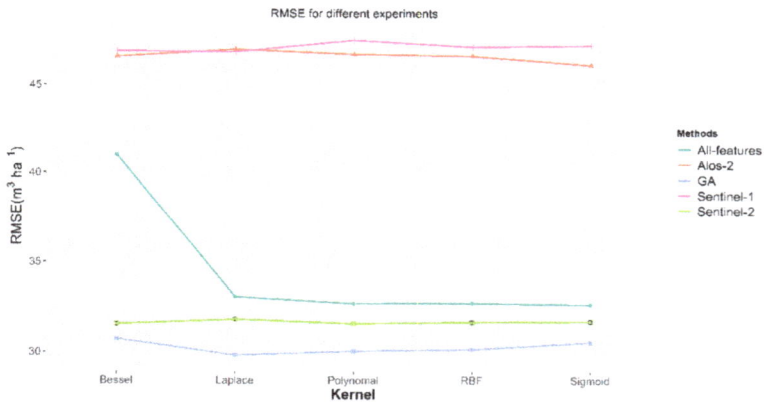

Figure 3. RMSE plot for different experiments.

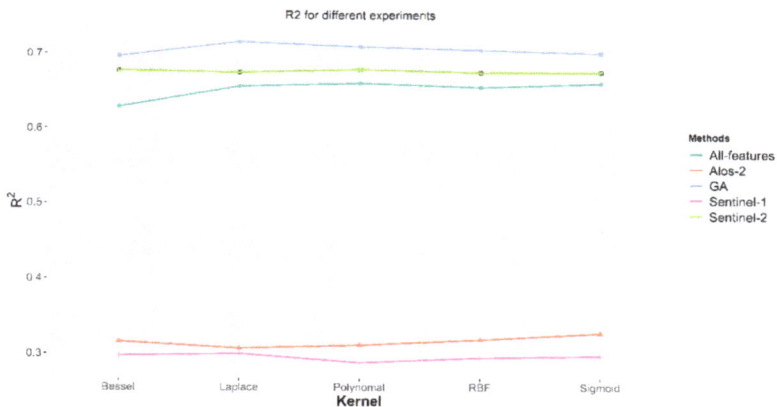

Figure 4. R^2 plot for different experiments.

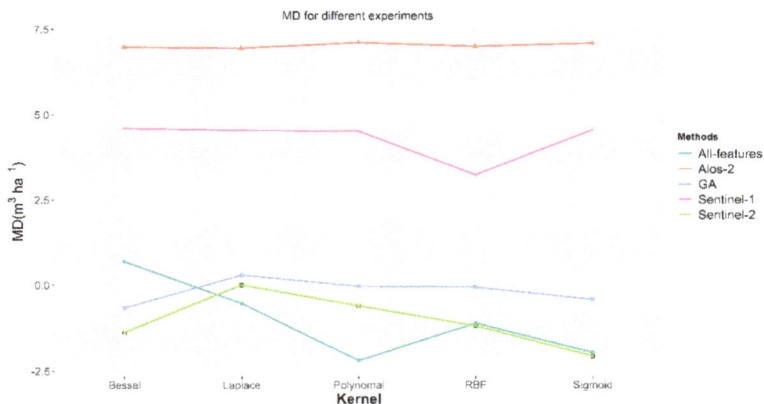

Figure 5. MD plot for different experiments.

3.1.2. ALOS-2

Here, 45 features were extracted from ALOS-2 data (see Table 4). The results are shown in Table 7.

Table 7. ALOS-2 modeling with 45 features.

Kernel	RMSE	R^2	MD	RMSE_SD	R^2_SD
Laplace	46.959	0.306	6.93	6.508	0.124
RBF	46.588	0.318	6.99	6.510	0.123
Sigmoid	46.086	0.327	7.09	6.543	0.124
Polynomial	46.691	0.311	7.1	6.583	0.121
Bessel	46.576	0.316	6.97	6.516	0.122

As shown in Table 7, the sigmoid kernel with provided a better performance compared to other kernels by returning RMSE = 46.086 and R^2 = 0.327. This was closely followed by the RBF kernel with RMSE = 46.588 and R^2 = 0.318. Concerning the stability, the RBF kernel was more stable than other kernels (RMSE_SD = 6.510 and R^2_SD = 0.123), followed by Bessel kernel. The Laplace kernel showed similar stability to that of Bessel kernel, yet a considerably lower performance compared with other kernels (see Table 7). The MD index shows that all of models significantly over estimate the GSV.

By considering both performance and stability and MD analysis, the RBF provided better results compared to the other kernels. However, all kernels performed roughly close to each other, with no significant difference between the kernels in the GSV modeling using the ALOS-2 features (see Figures 3–5).

3.1.3. Sentinel-1

In this experiment, seven features, extracted from Sentinel-1 data solely were used for the GSV modeling (see Table 3). The results are show in Table 8.

Table 8. Sentinel-1 modeling with seven features.

Kernel	RMSE	R^2	MD	RMSE_SD	R^2_SD
Laplace	46.815	0.299	4.536	6.360	0.117
RBF	47.093	0.294	3.232	6.349	0.111
Sigmoid	47.173	0.296	4.549	6.594	0.118
Polynomial	47.473	0.287	4.517	6.348	0.110
Bessel	46.879	0.297	4.596	6.302	0.113

As can be seen in Table 8, the Laplace kernel exceeded all other applied kernels with RMSE = 46.815 and R^2 = 0.299 again, closely followed by the Bessel kernel with RMSE = 46.879 and R^2 = 0.297. Moreover, the Bessel kernel provided more stable results than other kernels (RMSE_SD = 6.302 and R^2_SD = 0.113). The MD index shows that all of models significantly over estimate the GSV after ALOS2. By considering both performance and stability, the Bessel kernel was slightly better compared to the other kernels.

3.2. Multi-Sensor Models

Here, we employed 83 features, obtained from Sentinel-2, ALOS-2, and Sentinel-1 data (see Tables 2–4). The results are shown in Table 9.

Table 9. Performances achieved by modeling with 83 features.

Kernel	RMSE	R^2	MD	RMSE_SD	R^2_SD
Laplace	33.053	0.655	−0.538	3.431	0.071
RBF	32.701	0.654	−1.12	3.052	0.073
Sigmoid	32.636	0.659	−1.96	2.905	0.067
Polynomial	32.682	0.659	−2.19	2.884	0.070
Bessel	40.961	0.628	0.69	5.068	0.073

The sigmoid kernel performed best (RMSE = 32.636 and R^2 = 0.659), followed by polynomial kernel (RMSE = 32.682 and R^2 = 0.659). In addition the sigmoid kernel resulted in the most stable model with RMSE_SD = 2.905 and R^2_SD = 0.067. The MD index shows that all of models except Bessel moderately under estimate the GSV.

As can be observed in Figures 3–5, stacking the entire features into one feature set did not improve the results. This is mainly due to the joint effect caused by the curse of dimensionality and the limited sample size, and suggests the necessity of feature selection as an essential task.

3.3. Multi-Sensor Approach with Selected Feature

The Table 10 summarizes the results of models built with features optimized using the binary genetic algorithm as described in Section 2.4.2.

Table 10. Modeling optimized feature sets for each kernel.

Kernel	RMSE	R^2	MD	RMSE_SD	R^2_SD	Number of Features
Laplace	29.815	0.715	0.288	3.338	0.064	18
RBF	30.132	0.704	−0.069	2.595	0.060	19
Sigmoid	30.527	0.7	−0.43	2.903	0.058	16
Polynomial	30.034	0.708	−0.041	2.844	0.058	12
Bessel	30.707	0.696	−0.667	2.901	0.061	18

The Laplace kernel provided a superior predictive accuracy to other kernels. This was followed by the polynomial kernel. However, the RBF kernel proved to return the most stable predictions, followed by the polynomial kernel.

All in all, the highest accuracies for GSV prediction were returned by the polynomial kernel. Table 11 summarizes the range of various features selected for the GSV modeling using each kernel. The selected features mainly include the ALOS-2 polarimetric decomposition, Sentinel-1 H-A-Alpha decomposition, and Sentinel-2 spectral features including NIR, red edge, SWIR bands and vegetation indices. The proper feature selection which resulted in the parsimonious feature sets enables achieving both predictive accuracy and model stability and robustness. Also, the MD index for all of the models compared to other experiments approach zero, especially for the polynomial kernel (MD = −0.041).

As summarized in Figures 3–5 the performance of multi-sensor approach with selected features was slightly better than that achieved by Sentinel-2, which suggest that including selected polarimetry features along with the multi-spectral information improves the GSV across mountainous broadleaf stands. Also the over estimation or under estimation problem was solved in this experiment. For polynomial kernel the MD is equal to −0.041 and it is best MD index between all of experiments.

Table 11. Selected features for kernels by GA.

Kernel	Sensors	Selected Features by GA
Laplace	ALOS-2	Entropy, Anisotropy, Lambda2, Lambda3, Psi3, Alpha3
	Sentinel-1	Entropy, Anisotropy, Alpha
	Sentinel-2	B8, B5, B7, B8a, B11, B12, gndvi, pssra, tsavi
RBF	ALOS-2	Beta, Lambda, Freeman_dbl, Cloude_dbl, VanZyl_vol, Alpha2, Psi3, Alpha3
	Sentinel-1	Entropy
	Sentinel-2	B4, B5, B8a, B11, B12, gndvi, mcari, msavi2, pssra, tsavi
Sigmoid	ALOS-2	VanZyl_dbl, Psi1, Phi2, Psi3
	Sentinel-1	Entropy, Anisotropy
	Sentinel-2	B4, B5, B11, B12, gndvi, mcari, msavi2, pssra, s2rep, tsavi
Polynomial	ALOS-2	VanZyl_vol, Psi3, Alpha3
	Sentinel-1	Entropy
	Sentinel-2	B4, B5, B12, gemi, gndvi, mcari, pssra, tsavi
Bessel	ALOS-2	Entropy, Beta, VanZyl_dbl, Psi1, Phi1, Psi3, Phi3
	Sentinel-1	Entropy
	Sentinel-2	B4, B8, B5, B12, gemi, gndvi, mcari, pssra, s2rep, tsavi

4. Discussion

The main objective of this research was to estimate the GSV in heterogeneously-structured and mountainous Hyrcanian forests in northern Iran. Predictive models and continuous monitoring of forest are especially vital in Iran's Hyrcanian forests because of the current ongoing rate of degradation and their crucial role in Iran's Hyrcanian forest ecosystem. Thus, investigating the potential of multi-frequency SAR, multi-spectral optical data for generating the reliable predictive models are essential in this study area. In this research, the GSV modeling was carried out using the polarimetric and multi-spectral features and their combination using three different approaches. The SVR with five various kernels was used for modeling because of non-linear and complex relation of features with the GSV particularly in multi-sensor approach. Also, there were several challenges in this procedure including the severe topography, limited field samples, highly mixed tree species in various ages, and complex slope-aspect structure.

In the first experiment, the GSV was modeled using the Sentinel-2 multi-spectral features. B11 and B12 bands from SWIR spectral region showed a high contribution because of their sensitivity to the water content in the canopy, that confirmed in research by Chrysafis et al. (2017) [8] and Mura et al. (2018) [9]. We also found the NIR and the red edge bands as good features in modeling which is due to their sensitivity to the chlorophyll and pigments of tree leaf. Relevant works of Chrysafis et al. (2017) [8], Mura et al. (2018) [9], and Chrysafis et al. (2019) [62] also described these features as influential features for the GSV estimation in their research. Basically, the vegetation indices that use the NIR and red-edge spectral bands have an effective contribution in the GSV estimation. This is mainly because of that the VI uses the combination of spectral bands and reaches the information that cannot be extracted from the single spectral band, as Chrysafis et al. (2017) [8] found this in their research. The Mura et al. (2018) [9] and Chrysafis et al. (2019) [62] also found that VIs have an effective contribution to GSV estimation.

The performed analysis indicated that the GSV modeling with ALOS-2 polarimetric features produced non-satisfactory results. However, increasing the number of polarimetric features can improve the results. The two main challenges for GSV modeling with ALOS-2 data in this study area are the harsh topography and the limited number of field inventory samples. As shown in Figure 6, the study area embrace harsh mountainous forest stands with high elevational gradient and complex slope-aspect structure. In this type of topography, phenomena like shadow and layover occur in SAR data. Particularly, no scanned data from the study areas were found for shadowing. Nevertheless, the radiometric values were significantly affected in layover behavior. Another effect is the POA (see Section 2.3.3) which was compensated in the preprocessing steps and thus results in no notable problem for the GSV modeling [43]. Yet the effect of slope in range direction, known as angular variation effect (AVE), still remains. In areas with forest cover and longer radar wavelength, the AVE introduces a notable effect. As such the double scattering mechanism increases in the slopes that face the radar incoming wave, whereas it decreases in opposite slopes. In addition, the volume scattering mechanism increases since the severe topography causes an increase in cross-pol scattering. In a severe mountainous area like our research site, the proposed correction methods do not completely refine the results [42,63]. In addition, a further effect is the effective scattering area due to the non-homomorphic imagining of radar. This effect can be maximized in the mountainous areas and thus dramatically affect the geo-referencing of the applied SAR data. This effect causes the higher radiometric values in front slopes facing the incoming radar wave and lower radiometric values in opposite slopes [42,43,63].

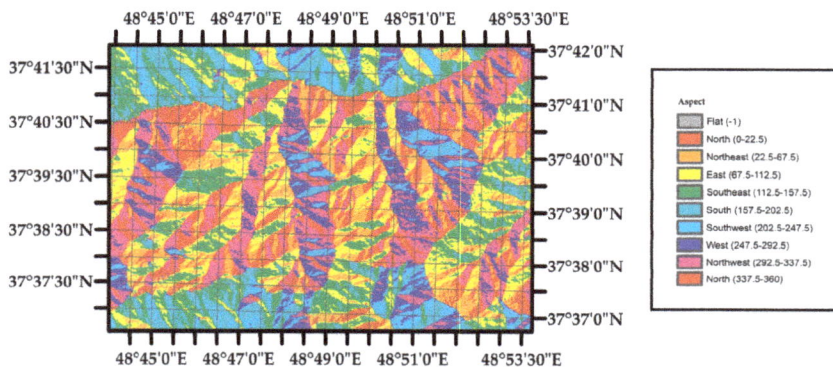

Figure 6. Aspect variation map in research area.

The second challenge for modeling GSV is caused by limited number of field inventory samples that is commonly the case across heterogeneously-structured uneven-aged Hyrcanian forests. Apart from statistical and model-related problems, this particularly hindered us from building species-specific GSV models and thus considering differences in scattering from tree species or different age classes. The collective effects of the above mentioned issues led to the rather mediocre results obtained by ALOS-2 data in our study (see Figures 3–5). The H-A-Alpha decomposition features have an average ability in the GSV estimation. Vafaei et al. (2018) [19] found that these features have an average ability for biomass estimation in Hyrcanian forests too. The Touzi decomposition's features are also effective, Sharifi et al. (2015) [64] found that the Tozi decomposition describes the target asymmetry in the forested area and can use for biomass estimation in Hyrcanian forest. The Van Zyle, Cloude, and Freeman decomposition features based on volume or double mechanism are effective in GSV modeling because of L-band wavelength penetration in the canopy and interaction with the tree's stem and ground. So they are reasonable features for the GSV modeling as Kumar et al. (2012) [65] found this behavior. The achieved performance and model stability are comparable to similar research. Compared with the only available relevant case study from the area, Vafaei et al. (2018) [19] used limited polarimetry features. Unlike their study, we extracted a large number of polarimetric features in our research. Especially comparing to Vafaei et al. (2018) [19] with ALOS-2 in the same region, his conclusion is that ALOS-2 has a very weak ability for biomass estimation ($R^2 = 16\%$). Since the biomass and GSV have near relation [5], we greatly improved the prediction ability for GSV with different target decomposition features ($R^2 = 32\%$).

The performed analysis indicated that the modeling with Sentinel-1 polarimetric features produced non-satisfactory results for GSV modeling. The Sentinel-1 features are not frequently used in forestry research because of the shorter wavelength (C-band) of the data. In this wavelength, the scattering only occurs in the upper part of the canopy. In our study, we applied the dual polarimetric H-A-Alpha decomposition. Our result has better performance (R^2 = 29%) than Mauya et al. (2019) [18] who only used the scattering coefficient in GRD mode and achieved R^2 = 18%. Therefore, we can conclude that using the Sentinel-1 dual polarimetry H-A-Alpha decomposition is useful for GSV estimation. As can be seen in Figures 3–5, the results obtained by the Sentinel-1 data were inferior than those of Sentinel-2 and ALOS-2 data. In addition to the underlying reasons as referred earlier, one may also note the lower penetration rate of the C-band Sentinel-1 compared with the L-band ALOS-2. This will result in a lower sensitivity to biophysical parameters such as GSV.

Finally, the performed analysis on the multi-sensor approach indicated that the GSV modeling with all of the features did not enhance the results compared to the modeling by Sentinel-2. This is mainly due to the curse of dimensionality effect and the limited sample size. To deal with this problem, we modeled the GSV by multi-sensor approach using the selected features. The results indicated that the modeling with the selected features provide better results than the modeling by Sentinel-2. The features were selected for each kernel separately proper feature selection procedure provides a good and stable GSV modeling which is not suffering from the curse of dimensionality anymore. Also, the over-estimation or under-estimation was reduced greatly by our feature selection method (see Figure 5). As a result, the simultaneous integration of multi-spectral and radar data features produce satisfactory and stable models for GSV modeling compare to the Mauya et al. (2019) [18], Vafaei et al. (2018) [19], and Sharifi et al. (2015) [64] works. Due to the general novelty of in-depth and state-of-the-art remote sensing analysis over Hyrcanian forests. In future research, issues such as the effects of different topographic corrections for polarimetry data, forest species type mapping, novel SVR kernels, application of feature fusion, and texture features will be explored.

5. Conclusions

This study generated empirical evidence on the use of ALOS-2, Sentinel-2, and Sentinel-1remotely sensing data for GSV estimation in Iran's Hyrcanian forests on a landscape scale. We used nu-SVR with five kernels for non-linear behavior modeling. The result showed that in single sensor approach only the Sentinel-2 returned better results, while shortcomings were observed when applying ALOS-2 and Sentinel-1 across heterogeneously-structured and mountainous Hyrcanian forests with limited field samples. In multi-sensor approach, the curse of dimensionality caused inferior results compared with the case of Sentinel-2. In addition, we tested selecting features in multi-sensor approach individually for each kernel by GA based on simultaneous optimization of diagnostic measures in a proposed fitness equation. This contributed to both higher performance and more model stability with the lowest underestimation.

Author Contributions: M.S.A., Y.M., H.L., and F.F. designed the research. F.F. provided field inventory sample data and advised the data analysis and interpretation. M.S.A. conducted data processing. M.S.A., Y.M., and H.L. analyzed the results. M.S.A. coded the statistical computation. M.S.A., Y.M., H.L., and F.F. wrote and commented the manuscript. H.L. and Y.M. were corresponding authors.

Funding: This research received no external funding.

Acknowledgments: We thank KNTU for technical support. We also thank ESA for providing open access Sentinel-1 and Sentinel-2 data. Also, JAXA is appreciated for providing ALOS-full polarimetry data via their data provision campaign.

Conflicts of Interest: The authors declare no conflict of interest.

References

1.   Ramezani, E.; Mohammad, R.; Mohadjer, M.R.M.; Knapp, K.-D.; Ahmadi, H.; Joosten, H. The late-Holocene vegetation history of the Central Caspian (Hyrcanian) forests of northern Iran. *Holocene* **2008**, *18*, 307–321. [CrossRef]

2. Attarchi, S.; Gloaguen, R. Classifying complex mountainous forests with L-band SAR and landsat data integration: A comparison among different machine learning methods in the hyrcanian forest. *Remote Sens.* **2014**, *6*, 3624–3647. [CrossRef]

3. United Nations Educational, Scientific and Cultural Organization (UNESCO), Hyrcanian Forests. Available online: https://whc.unesco.org/en/list/1584 (accessed on 5 July 2019).

4. Somogyi, Z.; Teobaldelli, M.; Federici, S.; Matteucci, G.; Pagliari, V.; Grassi, G.; Seufert, G. Allometric biomass and carbon factors database. *iForest Biogeosci. For.* **2008**, *1*, 107–113. [CrossRef]

5. West, P.W. *Tree and Forest Measurement*; Springer: New York, NY, USA, 2015.

6. Yadav, B.K.V.; Nandy, S. Mapping aboveground woody biomass using forest inventory, remote sensing and geostatistical techniques. *Environ. Monit. Assess.* **2015**, *187*, 308. [CrossRef] [PubMed]

7. Zhang, H.; Zhu, J.; Wang, C.; Lin, H.; Long, J.; Zhao, L.; Fu, H.; Liu, Z. Forest growing stock volume estimation in subtropical mountain areas using PALSAR-2 L-band PolSAR data. *Forests* **2019**, *10*, 276. [CrossRef]

8. Chrysafis, I.; Mallinis, G.; Siachalou, S.; Patias, P. Assessing the relationships between growing stock volume and Sentinel-2 imagery in a Mediterranean forest ecosystem. *Remote Sens. Lett.* **2017**, *8*, 508–517. [CrossRef]

9. Mura, M.; Bottalico, F.; Giannetti, F.; Bertani, R.; Giannini, R.; Mancini, M.; Orlandini, S.; Travaglini, D.; Chirivi, G. Exploiting the capabilities of the Sentinel-2 multi spectral instrument for predicting growing stock volume in forest ecosystems. *Int. J. Appl. Earth Obs. Geoinf.* **2018**, *66*, 126–134. [CrossRef]

10. Askne, J.I.H.; Fransson, J.E.S.; Santoro, M.; Soja, M.J.; Ulander, L.M.H. Model-based biomass estimation of a hemi-boreal forest from multitemporal TanDEM-X acquisitions. *Remote Sens.* **2013**, *5*, 5574–5597. [CrossRef]

11. Bharadwaj, P.S.; Kumar, S.; Kushwaha, S.P.S.; Bijker, W. Polarimetric scattering model for estimation of above ground biomass of multilayer vegetation using ALOS-PALSAR quad-pol data. *Phys. Chem. Earth* **2015**, *83–84*, 187–195. [CrossRef]

12. Ningthoujam, R.K.; Joshi, P.K.; Roy, P.S. Retrieval of forest biomass for tropical deciduous mixed forest using ALOS PALSAR mosaic imagery and field plot data. *Int. J. Appl. Earth Obs. Geoinf.* **2018**, *69*, 206–216. [CrossRef]

13. Gao, T.; Zhu, J.J.; Yan, Q.L.; Deng, S.Q.; Zheng, X.; Zhang, J.X.; Shang, G.D. Mapping growing stock volume and biomass carbon storage of larch plantations in Northeast China with L-band ALOS PALSAR backscatter mosaics. *Int. J. Remote Sens.* **2018**, *39*, 7978–7997. [CrossRef]

14. Antropov, O.; Rauste, Y.; Ahola, H.; Hame, T. Stand.-Level Stem Volume of Boreal Forests from Spaceborne SAR Imagery at L-Band. *IEEE J. Sel. Top. Appl. Earth Obs. Remote Sens.* **2013**, *6*, 35–44. [CrossRef]

15. Chowdhury, T.A.; Thiel, C.; Schmullius, C. Growing stock volume estimation from L-band ALOS PALSAR polarimetric coherence in Siberian forest. *Remote Sens. Environ.* **2014**, *155*, 129–144. [CrossRef]

16. Ghosh, S.M.; Behera, M.D. Aboveground biomass estimation using multi-sensor data synergy and machine learning algorithms in a dense tropical forest. *Appl. Geogr.* **2018**, *96*, 29–40. [CrossRef]

17. Laurin, G.V.; Pirotti, F.; Callegari, M.; Chen, Q.; Cuozzo, G.; Lingua, E.; Notarnicola, C.; Papale, D. Potential of ALOS2 and NDVI to estimate forest above-ground biomass, and comparison with lidar-derived estimates. *Remote Sens.* **2016**, *9*, 18. [CrossRef]

18. Mauya, E.W.; Koksinen, J.; Tegel, K.; Hamalainen, J.; Kauranne, T.; Kayhko, N. Modelling and predicting the growing stock volume in small-scale plantation forests of Tanzania using multi-sensor image synergy. *Forests* **2019**, *10*, 279. [CrossRef]

19. Vafaei, S.; Soosani, J.; Adeli, K.; Fadaei, H.; Naghavi, H.; Pham, T.D.; Bui, D.T. Improving accuracy estimation of forest aboveground biomass based on incorporation of ALOS-2 PALSAR-2 and sentinel-2A imagery and machine learning: A case study of the hyrcanian forest area (Iran). *Remote Sens.* **2018**, *10*, 172. [CrossRef]

20. The FWRO Technical Forestry Office. *The First Round of National Inventory of Hyrcanian Forests (2005–2007)*; The humid and semi-humid forests deputy of the FRWO; FRWO: Tehran, Iran, 2008.

21. European Space Agency (ESA). *Sentinel-2 User Handbook*, 1st ed.; European Space Agency (ESA): Paris, France, 2015; p. 64.

22. United States Geological Survey (USGS). SRTM. Available online: https://dds.cr.usgs.gov/srtm/ (accessed on 1 May 2019).

23. European Space Agency (ESA). *Sen2Cor Configuration and User Manual*; ESA Standard Document; ESA: Paris, France, 2018; Volume 2.5.5.

24. SNAP. European Space Agency (ESA), Ver 6.0. Available online: https://step.esa.int/main/ (accessed on 1 May 2019).

25. Pinty, B.; Verstraete, M.M. GEMI: A non-linear index to monitor global vegetation from satellites. *Vegetatio* **1992**, *101*, 15–20. [CrossRef]

26. Huete, A.R. A soil-adjusted vegetation index (SAVI). *Remote Sens. Environ.* **1988**, *25*, 295–309. [CrossRef]

27. Kaufman, Y.J.; Tanre, D. Atmospherically resistant vegetation index (ARVI) for EOS-MODIS. *IEEE Trans. Geosci. Remote Sens.* **1992**, *30*, 261–270. [CrossRef]

28. Baret, F.; Guyot, G.; Major, D.J. *TSAVI: A Vegetation Index Which Minimizes Soil Brightness Effects on LAI and APAR Estimation*; IEEE: Piscataway, NJ, USA, 1989.

29. Qi, J.; Kerr, Y.; Chehbouni, A. A modified soil adjusted vegetation index. *Remote Sens. Environ.* **1994**, *48*, 119–126.

30. Delegido, J.; Verrelst, J.; Meza, C.M.; Rivera, J.P.; Alonso, L.; Moreno, J. A red-edge spectral index for remote sensing estimation of green LAI over agroecosystems. *Eur. J. Agron.* **2013**, *46*, 42–52. [CrossRef]

31. Dash, J.; Curran, P.J. The MERIS terrestrial chlorophyll index. *Int. J. Remote Sens.* **2004**, *25*, 5403–5413. [CrossRef]

32. Musande, V.; Kumar, A.; Kale, K. Cotton crop discrimination using fuzzy classification approach. *J. Indian Soc. Remote Sens.* **2012**, *40*, 589–597.

33. Daughtry, C. Estimating corn leaf chlorophyll concentration from leaf and canopy reflectance. *Remote Sens. Environ.* **2000**, *74*, 229–239. [CrossRef]

34. Guyot, G.; Baret, F. Utilisation de la haute resolution spectrale pour suivre l'etat des couverts vegetaux. *Spectr. Signat. Objects Remote Sens.* **1988**, *287*, 279.

35. Crippen, R. Calculating the vegetation index faster. *Remote Sens. Environ.* **1990**, *34*, 71–73. [CrossRef]

36. Clevers, J.G.P.W.; Verhoef, W. LAI estimation by means of the WDVI: A sensitivity analysis with a combined PROSPECT-SAIL model. *Remote Sens. Rev.* **1993**, *7*, 43–64. [CrossRef]

37. Clevers, J.G.P.W.; de Jong, S.M.; Epema, G.F.; Addink, E.A. Meris and the Red-edge index. In *Second EARSeL Workshop on Imaging Spectroscopy*; EARSeL: Enschede, The Netherlands, 2000.

38. Blackburn, G.A. Quantifying Chlorophylls and Caroteniods at Leaf and Canopy Scales. *Remote Sens. Environ.* **1998**, *66*, 273–285. [CrossRef]

39. Gitelson, A.A.; Kaufman, Y.J.; Merzlyak, M.N. Use of a green channel in remote sensing of global vegetation from EOS-MODIS. *Remote Sens. Environ.* **1996**, *58*, 289–298. [CrossRef]

40. Pelich, R.; Lopez-Martinez, C.; Chini, M.; Hostache, R.; Matgen, P.; Ries, P.; Eides, G. Exploring dual-polarimetic descriptors for sentinel-L based ship detection. In Proceedings of the IGARSS 2018 IEEE International Geoscience and Remote Sensing Symposium, Valencia, Spain, 22–27 July 2018.

41. Lee, J.-S.; Pottier, E. *Polarimetric Radar Imaging: From BASICS to applications*; CRC Press: Boca Raton, FL, USA, 2009.

42. Villard, L.; Toan, T.L. Relating P-band SAR intensity to biomass for tropical dense forests in Hilly terrain: γ0ort0? *IEEE J. Select. Topics Appl. Earth Obs. Remote Sens.* **2015**, *8*, 214–223. [CrossRef]

43. Zhao, L.; Chen, E.; Li, Z.; Zhang, W.; Gu, X. Three-step semi-empirical radiometric terrain correction approach for PolSAR data applied to forested areas. *Remote Sens.* **2017**, *9*, 269. [CrossRef]

44. Lee, J.-S.; Lee, J.-S.; Wen, J.-H.; Ainsworth, T.L.; Chen, K.-S. Improved Sigma Filter for Speckle Filtering of SAR Imagery. *IEEE Trans. Geosci. Remote Sens.* **2009**, *47*, 202–213.

45. Mousavi, M.; Amini, J.; Maghsoudi, Y.; Arab, S. PolSAR speckle filtering techniques and their effects on classification. In Proceedings of the Imaging and Geospatial Technology Forum, IGTF 2015—ASPRS Annual Conference and Co-Located JACIE Workshop, Tampa, FL, USA, 4–8 May 2015.

46. Yamaguchi, Y.; Yajima, Y.; Yamada, H. A Four-Component Decomposition of POLSAR Images Based on the Coherency Matrix. *IEEE Geosci. Remote Sens. Lett.* **2006**, *3*, 292–296. [CrossRef]

47. Van Zyl, J.J. Application of Cloudes target decomposition theorem to polarimetric imaging radar data. In *Radar Polarimetry*; SPIE: Bellingham, WA, USA.

48. Freeman, A.; Durden, S.L. A three-component scattering model for polarimetric SAR data. *IEEE Trans. Geosci. Remote Sens.* **1998**, *36*, 963–973. [CrossRef]

49. Touzi, R. Target Scattering Decomposition in Terms of Roll-Invariant Target Parameters. *IEEE Trans. Geosci. Remote Sens.* **2007**, *45*, 73–84. [CrossRef]

50. Bishop, C.M. *Pattern Recognition and Machine Learning (Information Science and Statistics)*; Springer: New York, NY, USA, 2011.

51. Kuhn, M.; Johnson, K. *Applied Predictive Modeling*; Springer: New York, NY, USA, 2013.

52. Chang, C.-C.; Lin, C.-J. Training v-Support. Vector Regression: Theory and Algorithms. *Neural Comput.* **2002**, *14*, 1959–1977.

53. Smola, A.J.; Schölkopf, B. A tutorial on support vector regression. *Stat. Comput.* **2004**, *14*, 199–222. [CrossRef]

54. Karatzoglou, A.; Smola, A.; Hornik, K.; Zeileis, A. Kernlab-an S4 package for kernel methods in R. *J. Stat. Softw.* **2004**, *11*, 1–20. [CrossRef]

55. Jain, A.; Zongker, D. Feature selection: Evaluation, application, and small sample performance. *IEEE Trans. Pattern Anal. Machine Intell.* **1997**, *19*, 153–158. [CrossRef]

56. Jain, A.K.; Chandrasekaran, B. 39 Dimensionality and sample size considerations in pattern recognition practice. In *Handbook of Statistics*; Elsevier: Amsterdam, The Netherlands, 1982; pp. 835–855.

57. Whitley, D. A genetic algorithm tutorial. *Stat. Comput.* **1994**, *4*, 1–41. [CrossRef]

58. Yu, S.; Backer, S.D.; Scheunders, P. Genetic feature selection combined with composite fuzzy nearest neighbor classifiers for hyperspectral satellite imagery. *Pattern Recognit. Lett.* **2002**, *23*, 183–190. [CrossRef]

59. Li, Z.; Zan, Q.; Yang, Q.; Zhu, D.; Chen, Y.; Yu, S. Remote estimation of mangrove aboveground carbon stock at the species level using a low-cost unmanned aerial vehicle system. *Remote Sens.* **2019**, *11*, 1018. [CrossRef]

60. Yuan, H.; Yang, G.; Li, C.; Wang, Y.; Liu, J.; Yu, H.; Feng, H.; Xu, B.; Zhao, X.; Yang, X. Retrieving soybean leaf area index from unmanned aerial vehicle hyperspectral remote sensing: Analysis of RF, ANN, and SVM regression models. *Remote Sens.* **2017**, *9*, 309. [CrossRef]

61. Valbuena, R.; Hernando, A.; Manzanera, J.A.; Gorgens, E.B.; Almeida, D.R.A.; Mauro, F.; Garcia-Abril, A.; Coomes, D.A. Enhancing of accuracy assessment for forest above-ground biomass estimates obtained from remote sensing via hypothesis testing and overfitting evaluation. *Ecol. Model.* **2017**, *366*, 15–26. [CrossRef]

62. Chrysafis, I.; Mallinis, G.; Tsakiri, M.; Patias, P. Evaluation of single-date and multi-seasonal spatial and spectral information of Sentinel-2 imagery to assess growing stock volume of a Mediterranean forest. *Int. J. Appl. Earth Obs. Geoinf.* **2019**, *77*, 1–14. [CrossRef]

63. Small, D. Flattening Gamma: Radiometric Terrain Correction for SAR Imagery. *IEEE Trans. Geosci. Remote Sens.* **2011**, *49*, 3081–3093. [CrossRef]

64. Sharifi, A.; Amini, J. Forest biomass estimation using synthetic aperture radar polarimetric features. *J. Appl. Remote Sens.* **2015**, *9*, 097695. [CrossRef]

65. Kumar, S.; Pandey, U.; Kushwaha, S.P.; Chatterjee, R.S.; Bijker, W. Aboveground biomass estimation of tropical forest from Envisat advanced synthetic aperture radar data using modeling approach. *J. Appl. Remote Sens.* **2012**, *6*, 063588. [CrossRef]

Article

Fractional Cover Mapping of Invasive Plant Species by Combining Very High-Resolution Stereo and Multi-Sensor Multispectral Imageries

Siddhartha Khare [1,2,*], **Hooman Latifi** [3,4,*] , **Sergio Rossi** [1,5] **and Sanjay Kumar Ghosh** [2]

[1] Département des Sciences Fondamentales, Université du Québec à Chicoutimi,
   Saguenay, QC G7H 2B1, Canada
[2] Department of Civil Engineering, Geomatics Engineering Division, Indian Institute of Technology,
   Roorkee 247667, India
[3] Department of Photogrammetry and Remote Sensing, Faculty of Geodesy and Geomatics Engineering,
   K. N. Toosi University of Technology, Tehran 19967-15433, Iran
[4] Department of Remote Sensing, University of Würzburg, D-97074 Würzburg, Germany
[5] Key Laboratory of Vegetation Restoration and Management of Degraded Ecosystems,
   Guangdong Provincial Key Laboratory of Applied Botany, South China Botanical Garden,
   Chinese Academy of Sciences, Guangzhou 510650, China
* Correspondence: siddhartha.khare1@uqac.ca (S.K.); hooman.latifi@kntu.ac.ir (H.L.)

Received: 21 May 2019; Accepted: 26 June 2019; Published: 27 June 2019

Abstract: Invasive plant species are major threats to biodiversity. They can be identified and monitored by means of high spatial resolution remote sensing imagery. This study aimed to test the potential of multiple very high-resolution (VHR) optical multispectral and stereo imageries (VHRSI) at spatial resolutions of 1.5 and 5 m to quantify the presence of the invasive lantana (*Lantana camara* L.) and predict its distribution at large spatial scale using medium-resolution fractional cover analysis. We created initial training data for fractional cover analysis by classifying smaller extent VHR data (SPOT-6 and RapidEye) along with three dimensional (3D) VHRSI derived digital surface model (DSM) datasets. We modelled the statistical relationship between fractional cover and spectral reflectance for a VHR subset of the study area located in the Himalayan region of India, and finally predicted the fractional cover of lantana based on the spectral reflectance of Landsat-8 imagery of a larger spatial extent. We classified SPOT-6 and RapidEye data and used the outputs as training data to create continuous field layers of Landsat-8 imagery. The area outside the overlapping region was predicted by fractional cover analysis due to the larger extent of Landsat-8 imagery compared with VHR datasets. Results showed clear discrimination of understory lantana from upperstory vegetation with 87.38% (for SPOT-6), and 85.27% (for RapidEye) overall accuracy due to the presence of additional VHRSI derived DSM information. Independent validation for lantana fractional cover estimated root-mean-square errors (RMSE) of 11.8% (for RapidEye) and 7.22% (for SPOT-6), and R^2 values of 0.85 and 0.92 for RapidEye (5 m) and SPOT-6 (1.5 m), respectively. Results suggested an increase in predictive accuracy of lantana within forest areas along with increase in the spatial resolution for the same Landsat-8 imagery. The variance explained at 1.5 m spatial resolution to predict lantana was 64.37%, whereas it decreased by up to 37.96% in the case of 5 m spatial resolution data. This study revealed the high potential of combining small extent VHR and VHRSI- derived 3D optical data with larger extent, freely available satellite data for identification and mapping of invasive species in mountainous forests and remote regions.

Keywords: *Lantana camara*; SPOT-6; RapidEye; 3D; DSM; Fractional cover analysis

1. Introduction

Various studies have revealed the importance of plant biodiversity for the functioning of an ecosystem, which is closely connected with human activities [1,2]. However, apart from its significance for terrestrial ecosystems, it is much more important to emphasize the qualitative and quantitative changes or threats to plant biodiversity [3]. A proper regular and effective monitoring system is evidently required to characterize these changes. With the ability to view terrestrial vegetation from space, remote sensing has tremendous potential to provide long-term, continuous solutions at different spatial, temporal and spectral resolutions [4–6]. Satellite remote sensing datasets can represent a great opportunity when field-based observations are impossible or hampered [7–9]. High-resolution satellite data have enabled the development of species-level distribution maps along with three-dimensional (3D) information on forest structural traits such as species composition, canopy diameter, and distribution of age-classes [10]. In recent years, interest in employing multi-spectral, multi-temporal high and very high spatial resolution data to study biological invasions in plant communities has grown considerably [11].

Topographic information derived from remote sensing data enhances differentiation of plant or tree species that are spectrally similar. However, during the last decade, airborne laser scanning has been the primary data source to capture 3D information on forest vertical structure [12–14]. This tool is expensive, and possibly unavailable for developing countries [15]. In recent years, the improvement in space technology has reduced the gap in terms of spatial resolution (up to 30 cm ground sample distance) between aerial and satellite imagery [16]. Moreover, high-resolution digital surface models (DSM) are now more widely available through stereo imaging capacities and a worldwide access to very high-resolution (VHR) satellite data.

Over the last decade, satellite data for large areas provide information only on presence or absence of forests types or tree cover percentages. Prominently, AVHRR, MODIS or Landsat-based continuous (fractional) land-cover maps are available at coarser and medium spatial resolutions for forest cover change [17–19]. Other products also provide information, such as NDVI-derived vegetation fractional cover data of Oceansat-2 Ocean Color Monitor [20] for India and European initiative coordinated information on the environment (CORINE) derived land cover inventory data for coniferous and broadleaf tree species groups [21]. A review and comparison of various land cover products derived from satellite data can be found in [21]. These data provide an overview of forest types and species distribution at coarser (250 m to 1 km) and medium (30 m) spatial resolutions, but detailed information at finer spatial levels like plant species or trees is still lacking.

Previous studies analyzed detailed level species distribution for mapping and identification, utilizing VHR remote sensing data such as IKONOS or WorldView-2 (WV2) [22–27]. Similarly, a few studies with spatial resolution <5 m addressed mapping of invasive plant species using IKONOS [28], Cartosat-I [29] and Pléiades -1A [30] in the western Himalayan region of India. However, such studies can be only implemented in small areas, due to the high cost and scarcity of VHR and VHR stereo imagery (VHRSI) data compared with medium and coarser spatial resolution data. Thus, detailed studies covering larger geographic areas at higher spatial resolutions are still lacking [31], in particular across the high-altitude, species-rich Indian Himalayan region [32].

An alternative is to develop approaches to estimate the fraction of land cover within each optical pixel, by linking freely available medium spatial resolution (larger extent) with VHR (smaller extent) data. Fractional cover analysis has been subjected to intensive research using different satellite and aerial imageries. Recent studies combined VHR WV-2 with time series Landsat [33], IKONOS with Landsat [34,35] and GeoEye-1 with Landsat data [36] for generation of high spatial resolution fractional cover maps for larger geographical areas. It is thus practically crucial to utilize the combination of VHR with freely available medium spatial resolution datasets such as Landsat or MODIS due to their larger spatial coverage.

In this study, the potential of new-generation VHRSI and VHR multispectral satellite data was tested for classification and fractional cover analysis of an invasive shrub, lantana (*Lantana camara* L.), which extensively affects the western Himalayan forest in India. Fractional cover analysis was suggested to address the issue of estimating lantana distribution in Doon Valley. We aimed to (i) classify VHR SPOT-6 and RapidEye with additional 3D information from VHRSI SPOT-6 generated DSM; (ii) model the statistical relationship between fractional cover and spectral reflectance derived from high spatial resolution maps, and predict the fractional cover of lantana in a larger area based on the spectral reflectance of Landsat 8 imagery; and (iii) compare model performance between Landsat 8—SPOT-6 (1.5 m) and Landsat 8—RapidEye (5 m) fractional cover maps. We expected that Landsat-8-based upscaled lantana species information derived from input classified VHR imageries would provide an appropriate baseline for a fractional cover analysis approach for different types of forest regimes.

2. Materials and Methods

2.1. Study Area

The study area is located in the western Himalayan region of Doon valley, Uttarakhand, India (29,055' to 30,030' N and 77,035' to 78,024' E), at elevations ranging between 500 and 800 m above sea level (Figure 1). The climate is humid sub-tropical [37]. The temperature ranges between 16.7 and 36 °C during summer, and between 5.2 and 23.4 °C in winter [38]. Average annual rainfall is 2025 mm, and is mainly concentrated in the period between June and September. The Doon valley encompasses subtropical moist deciduous forests (MDF) dominated by sal trees (*Shorea robusta* G.) and *Mallotus philippensis* Lam., with *Clerodendrum infortunatum* L. and lantana in the understory. The Lachhiwala and Thano forest areas within the Doon valley were selected for this analysis due to availability of GPS observations of lantana locations in these forest sites.

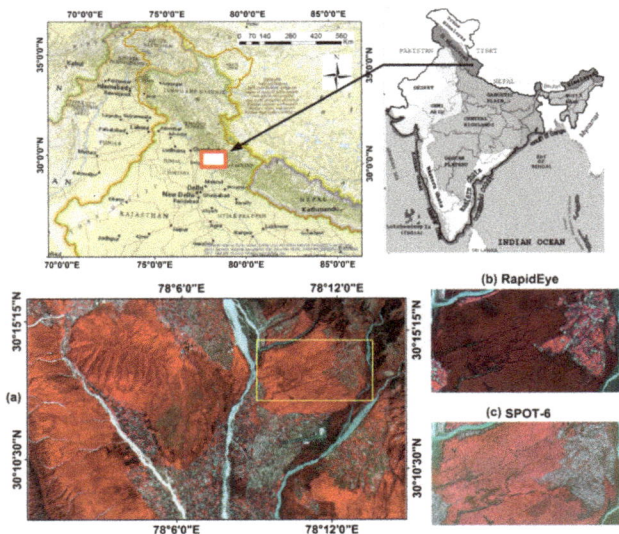

Figure 1. Study area location (**a**) Landsat-8 imagery with larger extent and yellow box showing the smaller extent of (**b**) RapidEye and (**c**) SPOT-6 multispectral imageries.

2.2. Satellite Data

We acquired satellite remote sensing data during April 2013, because shedding of leaves of overstory vegetation and visibility of understory vegetation (i.e., lantana) culminates in this month [39,40]. Orthorectified Level-3A, cloud free and pre-processed SPOT-6 imageries (multispectral and Panchromatic (PAN) stereo pair) and RapidEye (multispectral) were acquired (Table 1). The RapidEye sensor has an additional red-edge band with a spectral range of 690–730 nm; further details on the specifications of the RapidEye are given in [41]. We applied atmospheric corrections to convert DN values into surface reflectance using ATCOR 3 [42]. Selected vegetation indices (NDVI, MSAVI2 and NDRE) were generated from the surface reflectance values. Pre-processed Level-2, Landsat-8 OLI data were acquired from USGS (United States Geological Survey) for wider coverage (Figure 1). In this paper, four spectral bands (Blue, Green, Red and NIR) of Landsat-8 OLI were applied (Table 1).

Table 1. Details of satellite data used.

Satellite	Sensors	Date of Acquisitions	Spatial Resolution (m)
SPOT-6	(Stereo pair PAN)	5 April 2013	1.5
SPOT-6	(Blue, Green, Red and NIR)	25 April 2013	1.5
RapidEye	(Blue, Green, Red, Red-Edge and NIR)	12 April 2013	5
Landsat-8 OLI	(Blue, Green, Red and NIR)	11 April 2013	30

2.3. High-Resolution DSM

The DSM generation algorithm was adopted from [43]. The process comprised the following steps: tie points that were common to stereo pair images were generated using an automatic tie point generation tool. In the case of SPOT-6 imagery, RPC file was used as an input for tie point generation [44]. Tie points appearing within the overlap portion of the left and right images were identified. The resulting output consisted of the image location of tie points appearing within stereo pair rasters. Point cloud data were generated from stereo pairs using the image-matching point cloud generation algorithm eATE (Enhanced Automatic Terrain Extraction), which is an area-based method and uses a normalized cross correlation strategy [15,45] in Erdas Imagine (2015). The point cloud data were then interpolated into raster DSM [46,47] at a spatial resolution of 1.5 m by a triangulation technique implemented within ArcGIS (2015).

2.4. Image Classification and Generation of Reference Fractional Cover Data

Random forest (RF) classification was applied separately on both SPOT-6 and RapidEye multispectral VHR datasets (Figure 2). RF is a bootstrapped approach based on classification and regression trees (CARTs), with implications for both classification and predictive modelling. Detailed information on this algorithm and its application in remote sensing is available in [48,49]. The classification was carried out within the RSToolbox library [50] in R [51], following the steps as described by [52]. We generated 10-band stacked images separately for both RapidEye and SPOT-6 imageries along with the ancillary data and used them as input variables for the RF classifier. We stacked 5 spectral bands and 5 ancillary variables for RapidEye, and 4 spectral bands and 6 ancillary variables for SPOT-6.

Ancillary data (Table 2) in the case of classifying SPOT-6 image included (1) the 1.5 m spatial resolution generated DSM from SPOT-6 stereo pair, (2) slope and aspect calculated from the DSM, (3) texture measures of entropy and contrast derived for NIR and Red bands [53], (4) normalized difference vegetation index (NDVI), and (5) modified soil adjusted vegetation index2 (MASVI2). In the case of the RapidEye data, the ancillary data (Table 2) included (1) resampled 5-m DSM generated from the SPOT-6 stereo pair, (2) resampled slope and aspect calculated from the DSM, (3) texture measures of entropy and contrast derived for the NIR and red-edge bands [53], and (4) NDVI as well as (5) normalized difference red-edge index (NDRE).

Table 2. Global list of variables used as ancillary data.

Variables for Ancillary Data	Associated Satellite Data
DSM (1.5m)-elevation, slope and aspect	SPOT-6
Resampled DSM (5m)-elevation, slope and aspect	RapidEye
NDVI	SPOT-6 and RapidEye
MSAVI2	SPOT-6
NDRE	RapidEye
Texture measure (Entropy, Contrast)	SPOT-6 (NIR, Red bands) and RapidEye (NIR, Red-edge bands)

We followed straightforward rationales in selecting the vegetation indices used in our analysis. NDVI is a reliable proxy of photosynthetic activity and chlorophyll in vegetation [54]. NDRE leverages the information from Red-Edge region and is thus closely related to vegetation health [55] and nitrogen [56] content, and the MSAVI2 has been shown to account for soil background reflectance in different vegetation covers [57].

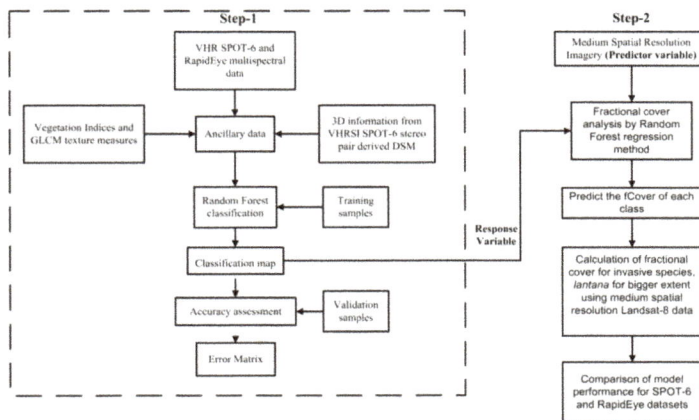

Figure 2. Workflow diagram for classification (Step-1) and fractional cover analysis (Step-2).

For each satellite data set, a total of 2000 random points were distributed throughout the training data from our input variables to grow 400 trees. For every tree, 66% of the data was used to construct the classification tree, while the remaining 33% was used for validation using out-of-bag (OOB) error [58]. OOB error was used from a classification accuracy matrix [59] as a tool to evaluate the final predictions with the withheld data and a variable importance plot to assess the relative importance of the 10 bands used in each classification [48]. Independence of the accuracy assessment was ensured by applying a 5-fold cross-validation with a 10% withhold of the training data [60], in addition to a post-classification accuracy assessment. Using this framework, the MDF and its surrounding locations were classified for SPOT-6 and RapidEye multispectral data. Post-classification accuracy assessment was conducted by generating stratified random points within each vegetation class. The points were manually attributed to their actual land-cover type through visual interpretation of Google Earth imagery [61] and to their attributed class by intersecting with the classified satellite image. A classification accuracy matrix was constructed to compare the reference class with the one assigned by the classifier and calculate the overall accuracy (OA), producer's and user's accuracies [62]. The Kappa coefficient was additionally calculated as a discrete multivariate statistic [62]. Lastly, variable importance as mean decrease in accuracy for each class was computed for classified RapidEye and SPOT-6 data. Mean decrease in accuracy measures the amount of mean standard error as decreased by removing a given input variable and it is calculated according to the increase in prediction error when OOB data (cases left out of the bootstrap sample) for that variable are permuted while all other variables are left unchanged [48].

2.5. Estimation of Fractional Cover for Larger Extent Using Landsat-8

Fractional cover maps of lantana for the Lachhiwala and Thano forests were generated by RF after creation of a reference dataset [33]. Fractional cover analysis was performed by taking the classified RapidEye and SPOT-6 imagery into account as predictors (Figure 2). We used the routine implemented within the RStoolbox [50]. Fractional cover extracts pixel values in a classified VHR image that corresponds to a random selection of medium spatial resolution pixels. It then calculates the percentage of classified image pixels that represent the cover type of interest. For example, with the 1.5 m pixel size of SPOT-6 VHR imagery and 30 m spatial resolution of Landsat-8, the sampling process uses 400 blocks of the 1.5-m resolution pixels corresponding to a single 30-m pixel and calculates the percentage of 1.5-m pixels that belong to the area covered by lantana. That is, with 100 pixels of lantana and 300 other class pixels, the value given for the output pixel would be 0.25, since 25% of the total amount of pixels belongs to lantana cover.

3. Results and Discussion

3.1. Classification of Combined Multispectral VHR and VHRSI DSM Data

VHRSI DSM is shown in Figure 3. Results of classifying SPOT-6 and RapidEye imageries in the MDF showed differences in cover estimation for all categories, in particular for sal tree, agriculture and lantana. This was related to the intrinsic difference in spatial resolutions (Figure 4). For example, both user accuracy (measure for commission) and producer accuracy (measure for omission) for tree-dominated classes (sal tree and lantana) were significantly higher for higher resolution SPOT-6 data, whereas lower resolution RapidEye data classification excessed in both measures for more homogeneously distributed agriculture class. In addition, a more contiguous and widely distributed lantana cover was estimated by SPOT-6. Overall, SPOT-6 and RapidEye showed 87.38 and 85.27% classification accuracy when tested by independent validation (Tables 3 and 4). Results were partially subjected to omission and commission errors. For lantana, 60 pixels in SPOT-6 and 48 pixels in RapidEye were assigned to shadow, probably because the larger sal trees were mainly growing on north-west slopes, and thereby casting shadows over the lantana cover (Tables 3 and 4). It was observed from previous ground and remote sensing-based studies that selected forest sites were mainly dominated by lantana as understory components [29,30,63]. In addition, lantana is an effective competitor with native plant species and is capable of interrupting the regeneration process of other indigenous species by reducing germination [63]. Therefore, GPS points and Google Earth VHR imagery were used to derive training data to portray shading of lantana by sal trees, aiming at a more realistic explanation of the lantana dominated cover. Furthermore, SPOT-6-based classification accuracy matrix revealed that 179 pixels related to agriculture were assigned to the built-up class, while 44 pixels of built-up were classified as agriculture. This is presumably related to the similarities in spectral signature characteristics of fallow land and built-up classes [30]. However, in the case of RapidEye, a negligible misclassification was observed between agriculture and built-up due to an additional red-edge band, which is able to precisely separate agriculture compared to the red band of RapidEye data [64].

Sal trees and lantana areas were occasionally misclassified (Tables 3 and 4). In SPOT-6 and RapidEye data, 224 and 350 pixels of lantana were assigned to sal trees class, respectively (Tables 3 and 4). These misclassified pixels mostly belonged to highly dense forest stands, thus the spectral signature and elevation from VHRSI DSM were not able to accurately discriminate lantana, which is mainly located in the understory, from sal tree stands. This resulted in cross-classification of lantana and sal tree classes. Lantana exhibited a kappa coefficient of 84.84% and 81.13%, whereas user's accuracy was 96.3% and 92.6% for SPOT-6 and RapidEye, respectively. VHR SPOT-6-based classification returned the second highest user's accuracy among all other classes compared to RapidEye when combined with very high spatial resolution DSM. In the case of RapidEye, user's accuracy of agriculture was the highest due to ability of the additional red-edge band to more precisely discriminate agriculture from sal trees and grasses. This confirmed the advantage of the applied classification for high spatial resolution multispectral data over medium and coarser spatial resolution satellite data. This is in agreement with former studies related to RF classification of VHR data such as WV-2 (2 m) for boreal forest habitats mapping [65], RapidEye (5 m) for classification of insect defoliation levels with help from the red-edge band, IKONOS (4 m) for tree health identification [66] and Pléiades-1B (2 m) for classification of wetland land-cover in arid regions [67].

Table 3. Classification accuracy matrix for SPOT-6.

Reference Data	Predicted Data						Row Total	Producer's Accuracy (%)	User's Accuracy (%)	Class Error (%)
	Built-Up	Agri-culture	Sal Tree	*Lantana*	Shadow	Water				
Built-up	945	44	0	1	0	11	1001	94.41	78.16	3.59
Agri-culture	179	806	8	1	7	0	1001	80.52	85.47	1.79
Sal tree	0	1	985	13	2	0	1001	98.40	80.94	3.67
Lantana	12	3	224	703	60	0	1002	70.16	96.30	1.68
Shadow	0	81	0	5	770	0	856	89.95	91.78	1.29
Water	73	8	0	7	0	913	1001	91.21	98.81	2.98
Column Total	1209	943	1217	730	839	924	5122			

Table 4. Classification accuracy matrix for RapidEye.

Reference Data	Predicted Data						Row Total	Producer's Accuracy (%)	User's Accuracy (%)	Class Error (%)
	Built-Up	Agri-culture	Sal Tree	*Lantana*	Shadow	Water				
Built-up	261	7	8	1	0	0	277	94.22	93.55	2.28
Agri-culture	4	522	48	1	0	7	582	89.69	97.57	2.29
Sal tree	7	3	935	37	19	0	1001	93.41	69.11	2.69
Lantana	0	3	350	601	48	0	1002	59.98	92.60	5.31
Shadow	0	0	12	9	53	0	74	71.62	41.09	13.60
Water	7	0	0	0	9	986	1002	98.40	99.30	2.88
Column Total	279	535	1353	649	129	993	3358			

The OOB error of SPOT-6, i.e., the internal consistency of the RF model for classifying the training data, varied between 1.3% and 3.7%, with an average OOB estimate of error of 2.5%. In addition, the applied 5-fold cross-validation resulted in 97.2 OA and 96.68% user's accuracy for lantana. A higher average OOB error of RapidEye was observed (3.58%) due to higher class error (13.6%) and lower user's accuracy (41.1%) of shadow (Table 4). The applied 5-fold cross-validation resulted in 96.47% OA and 94.67% user's accuracy for lantana. Overall, the classification of VHR SPOT-6 and RapidEye data achieved practically plausible results, which agreed with previous results of VHR data classification for detailed land cover (forests and plant species) mapping [27,33,68].

We used DSM (Figure 3) as an additional parameter in RF classification algorithm, which showed the variation in elevation from 400 to 1500 m. However, the elevation of Lachhiwala and Thano forests spans between 450 and 750 m (Figure 3). Since forest areas are heterogeneous and complex due to variations in tree height and stand density [69], high-resolution optical-based DSM can introduce different structural characteristics among understory and upperstory vegetation, which depends on the effects caused by tree orientation and sunlight [70]. In addition, shaded areas in high-density forest stands may reduce the quality of DSM due to variation in topography and sun angle [13]. One may also note the effect of seasonality on the quality of DSM [71]. We selected the SPOT scene within the leaf fall season of April, thus it resulted in partial extraction of ground and tree canopy by the applied image matching algorithm. Our results suggested that the classification accuracy increased by incorporating DSM from optical VHRSI SPOT-6 (at 1.5-m spatial resolution) for mapping understory (lantana) and upperstory (sal trees) vegetation. This was in agreement with previous studies, which highlighted the impact of elevation data as ancillary information in the RF classifier approach to differentiate spectrally similar objects and accuracy enhancement of land-cover mapping [72,73]. In addition, previous studies also reported that fine-scale DSM generated from VHRSI data may be used as an alternative to LiDAR data in areas with restricted accessibility [27,74]. Thus, very accurate large coverage DSM generation could be possible with the availability of stereo imaging capable VHR satellites and these may provide cost-effective solutions compared to expensive LiDAR technology for automatic classification of complex forest environments and delineation of understory plant species from upperstory vegetation.

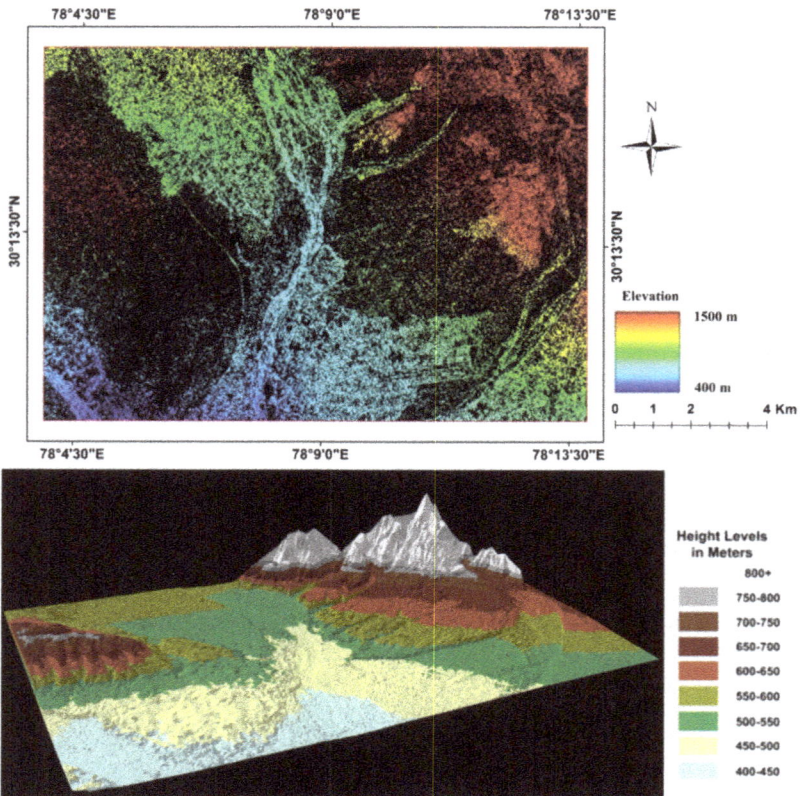

Figure 3. DSM of the study area represented by point cloud data (panel **a**) and 3D view (panel **b**). Black color in point cloud data represent NA values.

Overall, our results suggested the ability of SPOT-6 and RapidEye imageries to segregate lantana from dry sal trees and other land surface classes. However, the presence of lantana in forest areas generally increased confusion in the overall spectral signature within each pixel due to its competitive nature when growing with other vegetation types. Furthermore, comparisons across elevation gradients (see Figures 3b and 4) demonstrated that the classification of lantana and other land surface classes performed better in lower elevation areas (from 400 to 600 m a.s.l.).

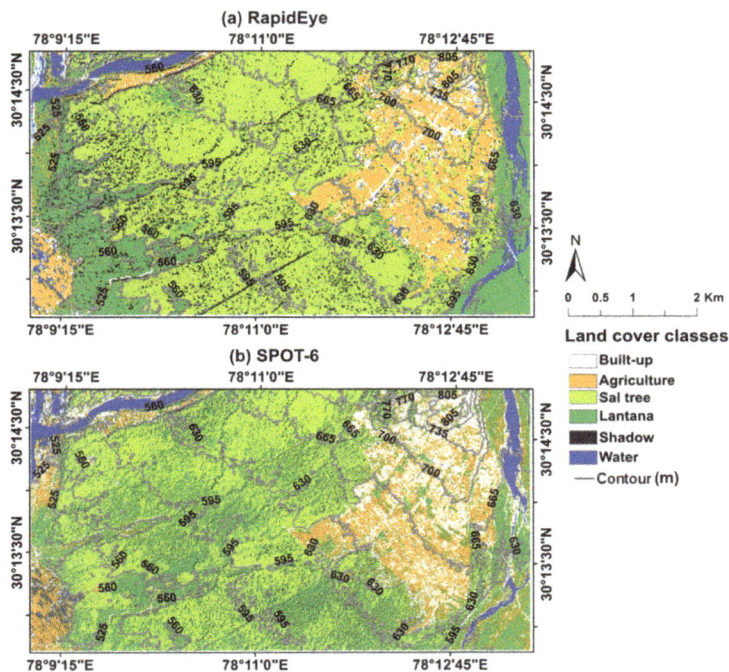

Figure 4. RF classification of (**a**) RapidEye and (**b**) SPOT-6 multispectral imageries for smaller extent.

3.2. Variable Importance

As expected from frequent field visits, DSM-derived elevation was among the most significant contributors to classification performance, although its influence was somewhat more pronounced when classifying RapidEye data. The overall importance of elevation was in line with a previous study [71] that applied Landsat multi-spectral data with ancillary data such as elevation (10 m contour interval), slope, and aspect for land-cover classification of mountainous areas. Accordingly, the DSM-derived elevation and slope combined with NDVI were the most important predictors for discriminating lantana (Figure 5). Whereas topographic information could be derived from a variety of 3D data sources, here we relied on those extracted from VHRSI due to a general absence of crown-penetrating LiDAR data across our study region. However, this was entirely due to a practical rationale; a recent study [74] also suggested that DSMs derived from VHRSI (WV-3 with 0.5 m spatial resolution) during leaf-off conditions were generally comparable to the LiDAR bare-earth DTM and may be used in land cover classification of vegetation during leaf-off seasons. In addition, red-edge spectral information of RapidEye data was ranked high when discriminating understory lantana from agriculture and built-up land cover, compared with texture parameters. Similar results were observed in a previous study that combined red-edge and texture parameters for paddy-rice crop classification [75]. Nevertheless, our study suggested that additional spectral information introduced by the red-edge band of RapidEye has the potential to discriminate invasive plants from other land-cover classes in relatively complex heterogeneous forest environments. However, future tests are suggested with freely available Sentinel-2 multispectral data featuring three red-edge bands.

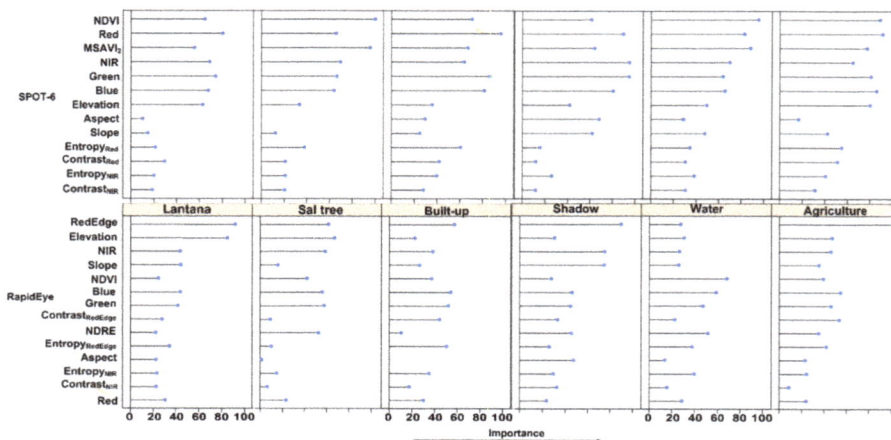

Figure 5. Variable importance shown as mean accuracy decrease (% decrease in overall accuracy) for SPOT-6 and RapidEye-based classification.

3.3. Landsat-8-Based Fractional Cover Maps for SPOT-6 and RapidEye

The RF models based on the results of SPOT-6 calibrated with larger extent Landsat-8-based predictors explained 64.38% of the variance for lantana, and 71.67% for sal trees. Moreover, 37.96% variance for lantana and 55.63% variance for sal trees were observed with RapidEye fractional reference map as an input (Table 5). The lower spatial resolution of Rapideye (5 m) compared with SPOT-6 (1.5 m) data resulted in lower observed variance when RapidEye was calibrated with larger extent Landsat-8. Furthermore, 40% fractional coverage of lantana was observed in lower elevation zones (400 to 500 m) from the classifications of Landsat-based upscaled maps (Figure 6) for both SPOT-6 and RapidEye fractional reference inputs. This result is in agreement with previous studies [30,63] that observed the dominance of lantana in open lowland (400–500 m a.s.l.) areas compared to the high-elevation (>500 m a.s.l.) sal-dominated areas.

Table 5. Fractional cover statistics for lantana and forest classes. Probability of significant relevance codes are $0.05 < p$: ***, $p < 0.001$.

Class	SPOT-6			RapidEye		
	R^2	RMSE (%)	Variance (%)	R^2	RMSE (%)	Variance (%)
Lantana	0.92 ***	7.22	64.38	0.85 ***	11.8	37.96
Sal trees	0.94 ***	7.73	71.67	0.86 ***	12.1	55.63

The most important variables for lantana estimation were the spectral reflectance of Landsat-8 (NIR, green and red) and April NDVI. Similar features were important for the sal trees models, with the main difference being the increased importance of NDVI. Higher R^2 and lower RMSE were observed for reference fractional cover data of SPOT-6 when compared to RapidEye for both lantana and sal trees (Table 5). The accuracy of fractional cover maps increased when higher spatial resolution maps (1.5 m) were used as a reference. This approach is in line with a previous study [33], which also suggested developing fractional cover maps for European spruce (*Picea abies* L.) and Scots pine (*Pinus sylvestris* L.) using RF regression by combining VHR WV-2(2 m) and medium spatial resolution Landsat time series data.

Our results were able to generate fractional cover maps in a heterogeneous environment when other classes were mixed with sal trees, especially soil-dominated and understory land covers such as lantana, agriculture and bare soil in the dry season. This was in agreement with an earlier study focusing on canopy fractional cover degradation mapping in a heterogeneous tropical forest of Brazil by combining Landsat ETM+ and VHR IKONOS data [35]. This research suggested the importance of reducing the mixed pixel problem in medium spatial resolution classification and the necessity of calibrating Landsat-derived results by the established model based on the combination of Landsat, SPOT-6, and RapidEye. The combination of open Landsat TM (30 m) and MODIS (250 m) were also used by [76] to map a forest and extract three fractions of vegetation, shadow and soil, to highlight deforestation. Since our approach included expensive VHR data, time series analysis could be limited by high cost. However, this type of analysis could be tested for a combination of open time series Sentinel-2 (10–20 m spatial resolution) and Landsat-8 (30 m spatial resolution) datasets.

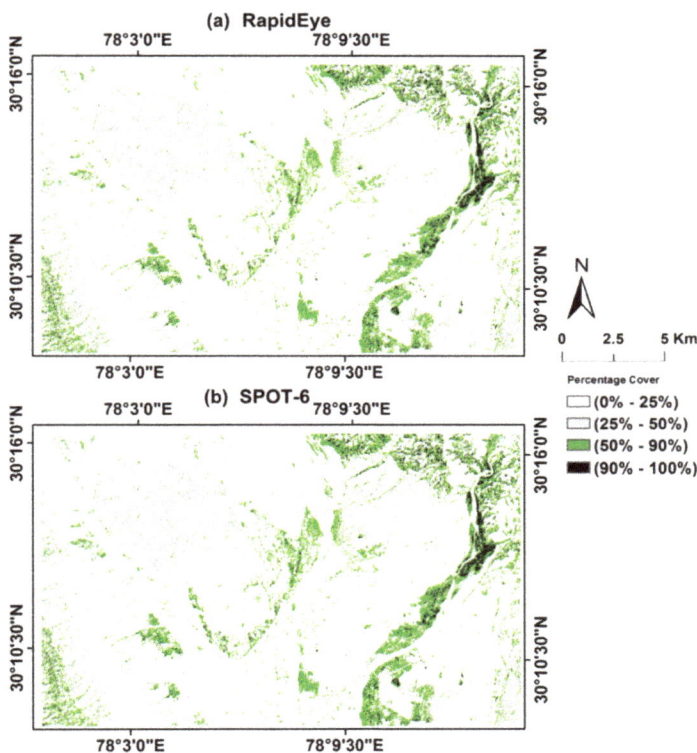

Figure 6. Landsat-8 classification for lantana within fractional cover thresholds (0%–25%, 25%–50%, 50%–90%, 90%–100%) for (**a**) RapidEye and (**b**) SPOT-6 data.

4. Conclusions

This study presented a fractional cover approach to predicting the proportion of lantana cover for a large area based on the spectral reflectance of medium spatial resolution multispectral Landsat-8 imagery in a Western Himalayan region of India. Training data for fractional cover analysis was classified with smaller extent VHR SPOT-6 (1.5 m) and RapidEye (5 m) imageries by adding VHRSI information derived from SPOT-6 data. Results of VHR maps showed 87.38% and 85.27% overall accuracy for SPOT-6 and RapidEye, respectively. Our observations suggested that 3D information from VHRSI optical satellite data played a crucial role in distinguishing understory (in our case lantana) from upperstory vegetation, being also a valid alternative to costly LiDAR data. We conclude that classification accuracy improves at increasing spatial resolution, with decreasing mixed pixel problems for fractional cover maps when spatializing data on larger geographical areas. This approach is consistent and reliable for large mountainous biodiversity hotspots, where direct field observations are prevented by harsh climatic conditions. This approach may be implemented for other species mapping over larger areas by combining freely available Sentinel-2 and Landsat datasets.

Author Contributions: S.K., H.L., and S.R. designed the research. S.K. and H.L. provided the satellite and other required data sources. S.K. conducted the data processing. S.K., H.L. and S.R. performed the analysis of the results and the statistical interpretation. S.K. performed the code programming, supported by H.L., S.K. and H.L. wrote the first draft of paper. H.L., S.R. and S.K.G. commented on the draft and all authors finalized it. H.L. and S.K. were corresponding authors.

Funding: The SPOT-6 multispectral and stereo pair dataset were funded by European Space Agency (ESA) with project id: 33429 and and RapidEye datasets were funded by RapidEye Science Archive (RESA) with project id: 00184.

Acknowledgments: We thank U.S. Geological Survey (USGS) EROS data center for Landsat-8 data, Department of Remote Sensing of University of Wuerzburg, Germany for providing a research residency to S. Khare, and A. Garside for checking the English text.

References

1. Duffy, J.E. Why biodiversity is important to the functioning of real-world ecosystems. *Front. Ecol. Environ.* **2009**, *7*, 437–444. [CrossRef]
2. Cardinale, B.J.; Duffy, J.E.; Gonzalez, A.; Hooper, D.U.; Perrings, C.; Venail, P.; Narwani, A.; Mace, G.M.; Tilman, D.; Wardle, D.A.; et al. Biodiversity loss and its impact on humanity. *Nature* **2012**, *486*, 59–67. [CrossRef]
3. Pasher, J.; Smith, P.A.; Forbes, M.R.; Duffe, J. Terrestrial ecosystem monitoring in Canada and the greater role for integrated earth observation. *Environ. Rev.* **2013**, *22*, 179–187. [CrossRef]
4. Turner, W.; Spector, S.; Gardiner, N.; Fladeland, M.; Sterling, E.; Steininger, M. Remote sensing for biodiversity science and conservation. *Trends Ecol. Evol.* **2003**, *18*, 306–314. [CrossRef]
5. Buchanan, G.M.; Nelson, A.; Mayaux, P.; Hartley, A.; Donald, P.F. Delivering a Global, Terrestrial, Biodiversity Observation System through Remote Sensing. *Conserv. Boil.* **2009**, *23*, 499–502. [CrossRef] [PubMed]
6. Leidner, A.K.; Brink, A.B.; Szantoi, Z. Leveraging Remote Sensing for Conservation Decision Making. *Eos Trans. Am. Geophys. Union* **2013**, *94*, 508. [CrossRef]
7. Nagendra, H.; Rocchini, D.; Ghate, R.; Sharma, B.; Pareeth, S. Assessing Plant Diversity in a Dry Tropical Forest: Comparing the Utility of Landsat and Ikonos Satellite Images. *Remote Sens.* **2010**, *2*, 478–496. [CrossRef]
8. Pettorelli, N.; Laurance, W.F.; O'Brien, T.G.; Wegmann, M.; Nagendra, H.; Turner, W. Satellite remote sensing for applied ecologists: Opportunities and challenges. *J. Appl. Ecol.* **2014**, *51*, 839–848. [CrossRef]
9. Lausch, A.; Bannehr, L.; Beckmann, M.; Boehm, C.; Feilhauer, H.; Hacker, J.; Heurich, M.; Jung, A.; Klenke, R.; Neumann, C.; et al. Linking Earth Observation and taxonomic, structural and functional biodiversity: Local to ecosystem perspectives. *Ecol. Indic.* **2016**, *70*, 317–339. [CrossRef]
10. Singh, J.S.; Roy, P.S.; Murthy, M.S.R.; Jha, C.S. Application of landscape ecology and remote sensing for assessment, monitoring and conservation of biodiversity. *J. Indian Soc. Remote Sens.* **2010**, *38*, 365–385. [CrossRef]
11. Joshi, P.K.; Rawat, G.S.; Padilya, H.; Roy, P.S. Biodiversity Characterization in Nubra Valley, Ladakh with Special Reference to Plant Resource Conservation and Bioprospecting. *Biodivers. Conserv.* **2006**, *15*, 4253–4270. [CrossRef]
12. Lim, K.; Treitz, P.; Wulder, M.; St-Onge, B.; Flood, M.; St-Onge, B. LiDAR remote sensing of forest structure. *Prog. Phys. Geogr. Earth Environ.* **2003**, *27*, 88–106. [CrossRef]
13. Latifi, H.; Heurich, M.; Hartig, F.; Müller, J.; Krzystek, P.; Jehl, H.; Dech, S. Estimating over- and understorey canopy density of temperate mixed stands by airborne LiDAR data. *Forestry* **2015**, *89*, 69–81. [CrossRef]
14. Vastaranta, M.; Wulder, M.A.; White, J.C.; Pekkarinen, A.; Tuominen, S.; Ginzler, C.; Kankare, V.; Holopainen, M.; Hyyppä, J.; Hyyppä, H. Airborne laser scanning and digital stereo imagery measures of forest

structure: Comparative results and implications to forest mapping and inventory update. *Can. J. Remote Sens.* **2013**, *39*, 382–395. [CrossRef]

15. Aguilar, M.A.; Saldana, M.D.M.; Aguilar, F.J. Generation and Quality Assessment of Stereo-Extracted DSM from GeoEye-1 and WorldView-2 Imagery. *IEEE Trans. Geosci. Remote Sens.* **2014**, *52*, 1259–1271. [CrossRef]

16. Piermattei, L.; Marty, M.; Karel, W.; Ressl, C.; Hollaus, M.; Ginzler, C.; Pfeifer, N. Impact of the Acquisition Geometry of Very High-Resolution Pléiades Imagery on the Accuracy of Canopy Height Models over Forested Alpine Regions. *Remote Sens.* **2018**, *10*, 1542. [CrossRef]

17. Zhu, Z.; Evans, D.L. US forest types and predicted percent forest cover from AVHRR data. *PE RS Photogramm. Eng. Remote Sens.* **1994**, *60*, 525–531.

18. Hansen, M.C.; Potapov, P.V.; Moore, R.; Hancher, M.; Turubanova, S.A.; Tyukavina, A.; Thau, D.; Stehman, S.V.; Goetz, S.J.; Loveland, T.R.; et al. High-Resolution Global Maps of 21st-Century Forest Cover Change. *Science* **2013**, *342*, 850–853. [CrossRef]

19. Kim, D.-H.; Sexton, J.O.; Noojipady, P.; Huang, C.; Anand, A.; Channan, S.; Feng, M.; Townshend, J.R. Global, Landsat-based forest-cover change from 1990 to 2000. *Remote Sens. Environ.* **2014**, *155*, 178–193. [CrossRef]

20. Kumar, A.S.; Radhika, T.; Saritha, P.; Keerthi, V.; Anjani, R.N.; Kumar, M.S.; Sekhar, K.S.; Satyanarayana, P.; Sudha, M.S.N.; Sai, M.S.; et al. Generation of Vegetation Fraction and Surface Albedo Products Over India from Ocean Colour Monitor (OCM) Data Onboard Oceansat-2. *J. Indian Soc. Remote Sens.* **2014**, *42*, 701–709. [CrossRef]

21. Pérez-Hoyos, A.; García-Haro, F.; San-Miguel-Ayanz, J. Conventional and fuzzy comparisons of large scale land cover products: Application to CORINE, GLC2000, MODIS and GlobCover in Europe. *ISPRS J. Photogramm. Remote Sens.* **2012**, *74*, 185–201. [CrossRef]

22. Carleer, A.P.; Wolff, E. Urban land cover multi-level region-based classification of VHR data by selecting relevant features. *Int. J. Remote Sens.* **2006**, *27*, 1035–1051. [CrossRef]

23. Mora, B.; Wulder, M.A.; White, J.C. Identifying leading species using tree crown metrics derived from very high spatial resolution imagery in a boreal forest environment. *Can. J. Remote Sens.* **2010**, *36*, 332–344. [CrossRef]

24. Kim, S.-R.; Lee, W.-K.; Kwak, D.-A.; Biging, G.S.; Gong, P.; Lee, J.-H.; Cho, H.-K. Forest Cover Classification by Optimal Segmentation of High Resolution Satellite Imagery. *Sensors* **2011**, *11*, 1943–1958. [CrossRef] [PubMed]

25. Immitzer, M.; Atzberger, C.; Koukal, T. Tree Species Classification with Random Forest Using Very High Spatial Resolution 8-Band WorldView-2 Satellite Data. *Remote Sens.* **2012**, *4*, 2661–2693. [CrossRef]

26. Waser, L.T.; Küchler, M.; Jütte, K.; Stampfer, T. Evaluating the Potential of WorldView-2 Data to Classify Tree Species and Different Levels of Ash Mortality. *Remote Sens.* **2014**, *6*, 4515–4545. [CrossRef]

27. Fassnacht, F.E.; Mangold, D.; Schäfer, J.; Immitzer, M.; Kattenborn, T.; Koch, B.; Latifi, H. Estimating stand density, biomass and tree species from very high resolution stereo-imagery—towards an all-in-one sensor for forestry applications? *For. Int. J. For. Res.* **2017**, *90*, 613–631. [CrossRef]

28. Kandwal, R.; Jeganathan, C.; Tolpekin, V.; Kushwaha, S.P.S. Discriminating the invasive species, 'Lantana' using vegetation indices. *J. Indian Soc. Remote Sens.* **2009**, *37*, 275–290. [CrossRef]

29. Kimothi, M.M.; Dasari, A. Methodology to map the spread of an invasive plant (*Lantana camara* L.) in forest ecosystems using Indian remote sensing satellite data. *Int. J. Remote Sens.* **2010**, *31*, 3273–3289. [CrossRef]

30. Khare, S.; Latifi, H.; Ghosh, S.K. Multi-scale assessment of invasive plant species diversity using Pléiades 1A, RapidEye and Landsat-8 data. *Geocarto Int.* **2018**, *33*, 681–698. [CrossRef]

31. Fassnacht, F.E.; Latifi, H.; Stereńczak, K.; Modzelewska, A.; Lefsky, M.; Waser, L.T.; Straub, C.; Ghosh, A. Review of studies on tree species classification from remotely sensed data. *Remote Sens. Environ.* **2016**, *186*, 64–87. [CrossRef]

32. Gairola, S.; Procheş, Ş.; Rocchini, D. High-resolution satellite remote sensing: A new frontier for biodiversity exploration in Indian Himalayan forests. *Int. J. Remote Sens.* **2013**, *34*, 2006–2022. [CrossRef]

33. Immitzer, M.; Böck, S.; Einzmann, K.; Vuolo, F.; Pinnel, N.; Wallner, A.; Atzberger, C. Fractional cover mapping of spruce and pine at 1 ha resolution combining very high and medium spatial resolution satellite imagery. *Remote Sens. Environ.* **2018**, *204*, 690–703. [CrossRef]

34. Metzler, J.W.; Sader, S.A. Model development and comparison to predict softwood and hardwood per cent cover using high and medium spatial resolution imagery. *Int. J. Remote Sens.* **2005**, *26*, 3749–3761. [CrossRef]

35. Wang, C.; Qi, J.; Cochrane, M. Assessment of tropical forest degradation with canopy fractional cover from Landsat ETM+ and IKONOS imagery. *Earth Interact.* **2005**, *9*, 1–18. [CrossRef]

36. Donmez, C.; Berberoglu, S.; Erdogan, M.A.; Tanriover, A.A.; Cilek, A. Response of the regression tree model to high resolution remote sensing data for predicting percent tree cover in a Mediterranean ecosystem. *Environ. Monit. Assess.* **2015**, *187*, 4. [CrossRef] [PubMed]

37. Champion, S.H.; Seth, S.K. *A Revised Survey of the Forest Types of India*; Manager of Publications: Delhi, Indian, 1968.

38. Peel, M.C.; Finlayson, B.L.; McMahon, T.A. Updated world map of the Köppen-Geiger climate classification. *Hydrol. Earth Syst. Sci.* **2007**, *11*, 1633–1644. [CrossRef]

39. Khare, S.; Latifi, H.; Ghosh, K. Phenology analysis of forest vegetation to environmental variables during pre-and post-monsoon seasons in western Himalayan region of India. *ISPRS Int. Arch. Photogramm. Remote Sens. Spat. Inf. Sci.* **2016**, *41*, 15–19. [CrossRef]

40. Khare, S.; Ghosh, S.K.; Latifi, H.; Vijay, S.; Dahms, T. Seasonal-based analysis of vegetation response to environmental variables in the mountainous forests of Western Himalaya using Landsat 8 data. *Int. J. Remote Sens.* **2017**, *38*, 4418–4442. [CrossRef]

41. Chander, G.; Haque, M.; Sampath, A.; Brunn, A.; Trosset, G.; Hoffmann, D.; Roloff, S.; Thiele, M.; Anderson, C. Radiometric and geometric assessment of data from the RapidEye constellation of satellites. *Int. J. Remote Sens.* **2013**, *34*, 5905–5925. [CrossRef]

42. Richter, R.; Center, R.S.D. *ATCOR: Atmospheric and Topographic Correction. DLR-German Aerospace Center*; Remote Sensing Data Center: Oberpfaffenhofen, Germany, 2004.

43. Kwoh, L.K.; Liew, S.C.; Xiong, Z. Automatic DEM generation from satellite image. In Proceedings of the 25th Asian Conference & 1st Asian Space Conference on Remote Sensing, Chiang Mai, Thailand, 22–25 November 2004; pp. 22–26.

44. Rottensteiner, F.; Weser, T.; Fraser, C.S. November. Georeferencing and orthoimage generation from long strips of ALOS imagery. In Proceedings of the 2nd ALOS PI Symposium, Rhodes, Greece, 3–7 November 2008.

45. Poon, J.; Fraser, C.S.; Chunsun, Z.; Li, Z.; Gruen, A. Quality Assessment of Digital Surface Models Generated From IKONOS Imagery. *Photogramm. Rec.* **2005**, *20*, 162–171. [CrossRef]

46. Toutin, T. Review article: Geometric processing of remote sensing images: Models, algorithms and methods. *Int. J. Remote Sens.* **2004**, *25*, 1893–1924. [CrossRef]

47. Krauß, T.; Reinartz, P.; Lehner, M.; Schroeder, M.; Stilla, U. DEM generation from very high resolution stereo satellite data in urban areas using dynamic programming. International Archives of the Photogrammetry. *Remote Sens. Spat. Inf. Sci.* **2005**, *36*, 1.

48. Breiman, L. Random forests. *Mach. Learn.* **2001**, *45*, 5–32. [CrossRef]

49. Belgiu, M.; Drăguţ, L. Random forest in remote sensing: A review of applications and future directions. *ISPRS J. Photogramm. Remote Sens.* **2016**, *114*, 24–31. [CrossRef]

50. Leutner, B.; Horning, N. *RStoolbox: Tools for Remote Sensing Data Analysis*; R Package Version 0.1; R Package Vignette: Madison, WI, USA, 2017; Volume 8.

51. Core Team, R.C. *R: A Language and Environment for Statistical Computing*; R Foundation for Statistical Computing: Vienna, Austria, 2013.

52. Wegmann, M.; Leutner, B.; Dech, S. (Eds.) *Remote Sensing and GIS for Ecologists: Using Open Source Software*; Pelagic Publishing Ltd.: Exeter, UK, 2016.

53. Zvoleff, A. *Glcm: Calculate Textures from Grey-Level Co-Occurrence Matrices GLCMs) in R*; R Package Version 1.0; R Package Vignette: Madison, WI, USA, 2015.

54. Gamon, J.A.; Kovalchuk, O.; Wong, C.Y.S.; Harris, A.; Garrity, S.R. Monitoring seasonal and diurnal changes in photosynthetic pigments with automated PRI and NDVI sensors. *Biogeosci. Discuss.* **2015**, *12*, 2947–2978. [CrossRef]

55. Modzelewska, A.; Stereńczak, K.; Mierczyk, M.; Maciuk, S.; Bałazy, R.; Zawiła-Niedźwiecki, T. Sensitivity of vegetation indices in relation to parameters of Norway spruce stands. *Folia For. Pol.* **2017**, *59*, 85–98. [CrossRef]

56. Cammarano, D.; Fitzgerald, G.J.; Casa, R.; Basso, B. Assessing the Robustness of Vegetation Indices to Estimate Wheat N in Mediterranean Environments. *Remote Sens.* **2014**, *6*, 2827–2844. [CrossRef]

57. Silleos, N.G.; Alexandridis, T.K.; Gitas, I.Z.; Perakis, K. Vegetation Indices: Advances Made in Biomass Estimation and Vegetation Monitoring in the Last 30 Years. *Geocarto Int.* **2006**, *21*, 21–28. [CrossRef]

58. Breiman, L. Bagging predictors. *Mach. Learn.* **1996**, *24*, 123–140. [CrossRef]

59. Provost, F.; Kohavi, R. Glossary of terms. *J. Mach. Learn.* **1998**, *30*, 271–274. [CrossRef]
60. Evans, J.S.; Murphy, M.A.; Holden, Z.A.; Cushman, S.A. Modeling species distribution and change using random forest. In *Predictive Species and Habitat Modeling in Landscape Ecology*; Springer: New York, NY, USA, 2011; pp. 139–159.
61. Fry, J.A.; Xian, G.; Jin, S.M.; Dewitz, J.A.; Homer, C.G.; Yang, L.M.; Barnes, C.A.; Herold, N.D.; Wickham, J.D. Completion of the 2006 national land cover database for the conterminous United States. *Photogramm. Eng. Remote Sens.* **2011**, *77*, 858–864.
62. Ismail, M.H.; Jusoff, K. Satellite data classification accuracy assessment based from reference dataset. *Int. J. Comput. Inf. Sci. Eng.* **2008**, *2*, 96–102.
63. Mandal, G.; Joshi, S.P. Eco-physiology and habitat invisibility of an invasive, tropical shrub (Lantana Camara) in western Himalayan forests of India. *For. Sci. Technol.* **2015**, *11*, 182–196.
64. Massetti, A. Assessing the effectiveness of RapidEye multispectral imagery for vegetation mapping in Madeira Island (Portugal). *Eur. J. Remote Sens.* **2016**, *49*, 643–672. [CrossRef]
65. Räsänen, A.; Rusanen, A.; Kuitunen, M.; Lensu, A. What makes segmentation good? A case study in boreal forest habitat mapping. *Int. J. Remote Sens.* **2013**, *34*, 8603–8627. [CrossRef]
66. Wang, H.; Zhao, Y.; Pu, R.; Zhang, Z. Mapping Robinia Pseudoacacia Forest Health Conditions by Using Combined Spectral, Spatial, and Textural Information Extracted from IKONOS Imagery and Random Forest Classifier. *Remote Sens.* **2015**, *7*, 9020–9044. [CrossRef]
67. Tian, S.; Zhang, X.; Tian, J.; Sun, Q. Random Forest Classification of Wetland Landcovers from Multi-Sensor Data in the Arid Region of Xinjiang, China. *Remote Sens.* **2016**, *8*, 954. [CrossRef]
68. Karlson, M.; Ostwald, M.; Reese, H.; Bazié, H.R.; Tankoano, B. Assessing the potential of multi-seasonal WorldView-2 imagery for mapping West African agroforestry tree species. *Int. J. Appl. Earth Obs. Geoinf.* **2016**, *50*, 80–88. [CrossRef]
69. Dash, J.P.; Watt, M.S.; Bhandari, S. Characterising forest structure using combinations of airborne laser scanning data, RapidEye satellite imagery and environmental variables. *Forestry* **2015**, *89*, 159–169. [CrossRef]
70. Qi, J.; Xie, D.; Yin, T.; Yan, G.; Gastellu-Etchegorry, J.-P.; Li, L.; Zhang, W.; Mu, X.; Norford, L.K. LESS: LargE-Scale remote sensing data and image simulation framework over heterogeneous 3D scenes. *Remote Sens. Environ.* **2019**, *221*, 695–706. [CrossRef]
71. Hyyppä, H.; Yu, X.; Hyyppä, J.; Kaartinen, H.; Kaasalainen, S.; Honkavaara, E.; Rönnholm, P. Factors affecting the quality of DTM generation in forested areas. *Int. Arch. Photogramm. Remote Sens. Spat. Inf. Sci.* **2005**, *36*, 85–90.
72. Gislason, P.O.; Benediktsson, J.A.; Sveinsson, J.R. Random Forests for land cover classification. *Pattern Recognit. Lett.* **2006**, *27*, 294–300. [CrossRef]
73. Corcoran, J.M.; Knight, J.F.; Gallant, A.L. Influence of Multi-Source and Multi-Temporal Remotely Sensed and Ancillary Data on the Accuracy of Random Forest Classification of Wetlands in Northern Minnesota. *Remote Sens.* **2013**, *5*, 3212–3238. [CrossRef]
74. DeWitt, J.D.; Warner, T.A.; Chirico, P.G.; Bergstresser, S.E. Creating high-resolution bare-earth digital elevation models (DEMs) from stereo imagery in an area of densely vegetated deciduous forest using combinations of procedures designed for lidar point cloud filtering. *GIScience Remote Sens.* **2017**, *54*, 1–21. [CrossRef]
75. Kim, H.-O.; Yeom, J.-M. Effect of red-edge and texture features for object-based paddy rice crop classification using RapidEye multi-spectral satellite image data. *Int. J. Remote Sens.* **2014**, *35*, 1–23. [CrossRef]
76. Lu, D.; Batistella, M.; Moran, E.; Hetrick, S.; Alves, D.; Brondizio, E. Fractional forest cover mapping in the Brazilian Amazon with a combination of MODIS and TM images. *Int. J. Remote Sens.* **2011**, *32*, 7131–7149. [CrossRef]

![forests logo] *forests*

MDPI

Article

Relationships between Satellite-Based Spectral Burned Ratios and Terrestrial Laser Scanning

Akira Kato [1],*, L. Monika Moskal [2], Jonathan L. Batchelor [2], David Thau [3] and Andrew T. Hudak [4]

[1] Graduate School of Horticulture, Chiba University, Chiba 2710092, Japan
[2] Precision Forestry Cooperative, School of Environmental and Forest Sciences, College of the Environment, University of Washington, Seattle, WA 98118, USA; lmmoskal@uw.edu (L.M.M.); jonbatch@uw.edu (J.L.B.)
[3] World Wildlife Fund, Washington, DC 20037, USA; thau@wwf.org
[4] Rocky Mountain Research Station, USDA Forest Service, Moscow, ID 83843, USA; ahudak@fs.fed.us
* Correspondence: akiran@faculty.chiba-u.jp; Tel.: +81-47-308-8892

Received: 5 April 2019; Accepted: 18 May 2019; Published: 23 May 2019

check for updates

Abstract: Three-dimensional point data acquired by Terrestrial Lidar Scanning (TLS) is used as ground observation in comparisons with fire severity indices computed from Landsat satellite multi-temporal images through Google Earth Engine (GEE). Forest fires are measured by the extent and severity of fire. Current methods of assessing fire severity are limited to on-site visual inspection or the use of satellite and aerial images to quantify severity over larger areas. On the ground, assessment of fire severity is influenced by the observers' knowledge of the local ecosystem and ability to accurately assess several forest structure measurements. The objective of this study is to introduce TLS to validate spectral burned ratios obtained from Landsat images. The spectral change was obtained by an image compositing technique through GEE. The 32 plots were collected using TLS in Wood Buffalo National Park, Canada. TLS-generated 3D points were converted to voxels and the counted voxels were compared in four height strata. There was a negative linear relationship between spectral indices and counted voxels in the height strata between 1 to 5 m to produce R^2 value of 0.45 and 0.47 for unburned plots and a non-linear relationship in the height strata between 0 to 0.5m for burned plots to produce R^2 value of 0.56 and 0.59. Shrub or stand development was related with the spectral indices at unburned plots, and vegetation recovery in the ground surface was related at burned plots. As TLS systems become more cost efficient and portable, techniques used in this study will be useful to produce objective assessments of structure measurements for fire refugia and ecological response after a fire. TLS is especially useful for the quick ground assessments which are needed for forest fire applications.

Keywords: forest fire; google earth engine; terrestrial laser scanner; laser; ground validation

1. Introduction

The area burned by a fire and the severity are two key descriptors of forest fires. Fire severity is directly related to the amount of vegetation consumed by fire, and the regeneration rates of vegetation after a fire [1]. This removal of vegetation is a contributing factor to post fire erosion [2]. Quantifying fire severity can be difficult, and improving the accuracy of the assessment will aid in post fire restoration efforts. Current methods of assessing fire severity are usually limited to on-site visual inspection of a post fire landscape [3], or the use of satellite and aerial images to quantify severity over larger areas. For satellite image analysis, the Normalized Burn Ratio (NBR) is a common spectral index used to assess fire severity. NBR uses the near-infrared and mid-infrared spectral regions to create a normalized index.

$$NBR = (\rho MIR - \rho NIR)/(\rho MIR + \rho NIR), \tag{1}$$

where ρMIR is the mid-infrared and ρNIR is the near-infrared spectral reflectance band.

NBR is widely utilized in fire monitoring protocols and is used to determine the extent of the area burned [4,5]. The difference between pre- and post-fire NBR (differential NBR, dNBR) has been shown to be more effective at describing fire severity than the differential Normalized Difference Vegetation Index (dNDVI) [6]. Related to dNBR is the Relative differenced NBR (RdNBR). Miller and Thode [7] proposed RdNBR as an improved version of dNBR.

$$dNBR = (NBRprefire - NBRpostfire), \tag{2}$$

$$RdNBR = dNBR/\sqrt{ABS(NBR_{prefire})}, \tag{3}$$

where ρMIR is the mid-infrared, ρNIR is the near-infrared, and ρRed is the red spectral reflectance band. The NBR ranges from −1 to 1. dNBR ranges from −2 to 2. A high value of dNBR and RdNBR indicates high fire severity.

Which index is better at describing fire severity is dependent on several factors [1,8]. The deviation of the indices from on the ground assessments of fire severity can be related to seasonal and topographic conditions [9]. In such instances, the indices can have a limited capability to accurately characterize fire severity [10–12].

On the ground, one of the most widely used methods to characterize fire severity is the Composite Burn Index (CBI, [3]). The CBI is mainly derived from visual estimation and subjective judgement and is a simple and fast approach. However, CBI is prone to observer bias. The values are influenced by the observers' knowledge of the local ecosystem and ability to accurately assess several forest structure measurements [13]. It is still unknown how structural components of the post fire forest influence the spectral signatures detected by satellites, as well as the fact that CBI assigns a severity value from both overstory and understory vegetation assessment.

To model the structural components of a post fire forest in relation to RdNBR, Miller and co-authors [14] used the Forest Vegetation Simulator (FVS, [15]). The FVS was used with data from the US Forest Service's Forest Inventory and Analysis (FIA) to simulate forest structure attributes for a range of variables. FVS was accurate in modeling resultant forest structure attributes for high severity fires, but failed in moderate and low severity burns. There is a need for high-resolution quantification of three-dimensional forest structure in relation to fire severity. For this study, a terrestrial laser scanner is used to produce objective and highly detailed 3D scans of forest structure at various levels of fire severity.

Three-dimensional point clouds acquired through TLS have been used in numerous ecosystem studies, including tree stem reconstruction [16,17], measuring biomass of saplings [18], determining leaf angle and distribution [19], quantifying canopy gaps [20], and various other ecological applications [21]. Fine scale measurements from TLS have also been used for forest inventory [22], biomass allometry [23], and tree species identification [24]. Using multi-temporal TLS data, small changes in tree height have been quantified [25], as well as spring phenology [26], crown competition [27], and biomass [28]. TLS has also been used previously to detect forest structural change caused by fire [29]. TLS has proven to be a valuable tool in not only quantifying above-ground biomass and structure, but also measuring fine scale change.

The objective of this study is to utilize terrestrial lidar to quantify and compare forest structure attributes with the spectral signatures of Landsat satellite images. This study notably proposes a technique to compare the forest structure in different height strata with burned and unburned spectral reflectance from satellite images.

2. Materials and Methods

2.1. Study Site

The study site was located at Wood Buffalo National Park (WBNP, Figure 1) in Northwest Territories and Alberta, Canada. WBNP became a national park in 1922 and is a designated wildlife refuge. Lightning is the major source of forest fires [30,31], and the park is located within the area known as the fire hot spot of Canada [32,33]. The dominant tree species are Jack pine (*Pinus banksiana* Lamb.), aspen (*Populus tremuloides* Michx.), balsam poplar (*Populus balsamifera* L.), white spruce (*Picea glauca* (Moench) Voss), black spruce (*Picea Mariana* (Mill.) BSP), and tamarack (*Larix laricana* (Du Roi) K. Koch).

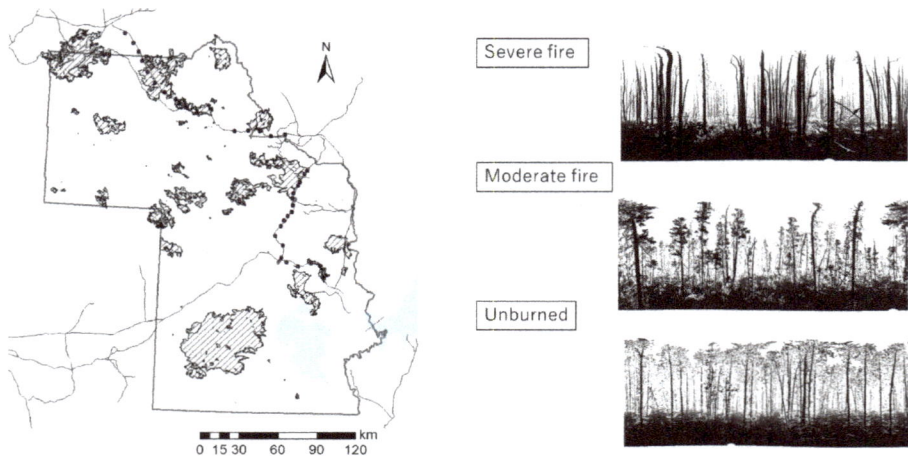

Figure 1. The study site with Terrestrial Lidar Scanning (TLS) locations denoted by ●. The thin black lines are roads and the shaded polygons are recently burned areas (within the last 3 years). To the right, panoramic views of TLS point clouds of different fire conditions.

2.2. Methodology

Species composition and stand density are major factors that influence the spectral signature of an area. A change in the spectral reflectance observed by a satellite between either two different sites, or the same site at different times, is likely due to differences in species composition or stand density. The area of our study, i.e., within the boreal forest biome, is covered mainly by homogeneous stands of the same age and species. The spectral signature of the area is thus also mainly homogenous, unless altered by some disturbance event (e.g., fire). The area largely consists of native forest stands that have been self-thinning and undergoing a natural succession process [34,35]. Forest succession entails a relatively simple structural change, and stand development can be related to the spectral change.

The spectral change due to fire is derived from pre- and post-fire Landsat image analysis using the Google Earth Engine platform. The structural components of the forest are measured using a Terrestrial Laser Scanner (TLS) to generate a 3D model of the area. The scans were taken in September of 2016.

To identify the spectral change, both dNBR and RdNBR are calculated. Our area of interest covered 44,807 km^2 requiring several Landsat image tiles to cover the extent. We utilize Google Earth Engine (GEE) to visualize and analyze the satellite imagery and use the GEE servers to compute dNBR and RdNBR. The use of an online platform greatly speeds up the processing and spares us the onerous task of downloading and processing gigabytes of image data [36]. To compute dNBR and RdNBR, pre- and post-fire dates need to be identified. However, the fires happened in different times and places within our 44,807 km^2 study site. To identify the location and date of each fire, a composite imaging technique is used to make pre- and post-fire images. The composite image technique uses a stack of time-series images to select and store the pixel value when the criteria are matched. In this study, NBR values are stored when NDVI is maximum for pre-fire condition and NBR is minimum for post-fire condition. This process is conducted within the study area using multi-temporal Landsat 8 images with a 16-day revisit cycle during the year of 2016. The images are only selected during the fire season from June 1st to October 1st. Using this date range, we avoid erroneous values caused by snow on the ground in spring and winter. An unburnt area has a high NDVI value while the same area post fire has a low NBR value. The difference between those values is used to produce dNBR and RdNBR values for each pixel.

The dNBR and RdNBR values are compared to structure measurements obtained with TLS. The portable laser sensor TX5 (Trimble Inc., Sunnyvale, CA, USA) is used for data collection. The 3D scan data has the potential to quantify the structural differences more accurately between unburned and recently-burned sites, compared to human visual observation. The position of the scan locations is recorded with a consumer grade GPS unit (Montana 600, Garmin Inc., Olathe, KS, USA) with a location error of 5 m. This locational error is far less than the half pixel size of 30 m Landsat imagery. The TX5 TLS system uses phase shift detection to generate 3D points and the horizontal and vertical scan line resolution is set to 0.0167°.

Within our study site, there is a road network bisecting the park in north/south and east/west directions. The TLS locations are chosen along this road network for ease of data collection (Figure 1). A systematic random sampling method is used to select sites to cover the diverse range of tree height and species within the large WBNP area. The scanner collects data every 10 km along the road (black dots along the road in Figure 1). The scanning locations are at least 30 m away from the road (perpendicular to the angle along the road) to insure the scan location is within a Landsat pixel that does not include reflectance from the road. The total 32 TLS scans were collected during Sept. 2016 and covered the diverse structural range from unburned sites to sites with high severity burn (Figure 1). All TLS scanning locations ($n = 32$) are divided into two groups: unburned plots (no fire since 1981, $n = 21$) and recently burned plots (most recent fire within the last 3 years, $n = 11$). Historical fire dates are determined using a fire map provided by the Canadian Park Services.

A multi-step process is used to derive structural metrics from the 3D TLS point clouds (Figure 2). A Digital Terrain Model (DTM) is created from the raw 3D point cloud using the climbing and sliding method [37]. The height of the points is normalized using the derived DTM. Only points within a 15 m radius of the scan position are used. A 15 m radius provided a point density sufficient to generate DTMs and is the plot size suggested by the CBI field protocol. The CBI field protocol includes cover change estimation, species identification, dead or live judgement, color change assessment, and soil disturbance estimation. They are not directly comparable to the number of voxels. However, the 3D data can contribute structure measurements by height strata (Box 1). The structure assessment is a major factor to determine CBI values. Therefore, this study is not aimed to compare values between CBI and TLS 3D data. The CBI data needs to be collected twice: right after fire for Initial Assessment (IA) and one year after fire for Extended Assessment (EA). Our field samplings were not collected right after fire. The timing to take our data is different from CBI.

To quantify the vertical forest structure of the scan sites, voxels are created at a 0.25 m resolution. The voxel size is fitted to the average stem diameter of this study site. A voxel is created when there is a laser return point located within a 0.25 m^3 grid cell. The number of voxels at different height strata is counted. There are four height strata adapted to this voxel analysis from the original CBI definition: 0 m to 0.5 m (ground surface), 0.5 m to 1m (shrubs), 1 m to 5 m (shrubs and understory), and 5 m above (overstory trees). The number of voxels at each stratum is correlated with the spectral reflectance measured by Landsat. The voxel is used rather than counting the number of raw returns, because the voxel approach can reduce the bias caused by the distance from the sensor and normalizes point density as points are inherently dense and close to the TLS sensor.

The relationship between counted voxels and spectral indices is examined at plot and strata levels. The total voxel count per plot is used to obtain the general relationship. The counts are divided into strata to determine which strata are more significantly related with the spectral change. For the specific strata, the stronger or strongest correlation is found. A linear or non-linear (2nd order polynomial) equation is fitted to the distribution between counted voxels and spectral indices. Both equations used in this study are assessed with R^2 values to indicate how strong the correlations are. Moreover, hypothesis tests are applied to test the difference between the linear and non-linear case. The *t*-test is used for the linear case and Shapiro-Wilk normality test is used for the goodness of fit for the non-linear case. Both cases used a significance level of 0.05. The statistical tests are indicated by $p < 0.05$ (a significant *t*-test result) for the linear case and $p > 0.05$ (a not-significant result with Shapiro-Wilk normality) for the non-linear case.

Box 1. Comparison between Composite Burn Index (CBI) and Terrestrial Lidar Scanning (TLS) measurements.

List of CBI measurement (ocular estimation and judgement)	List of TLS measurement for this study
A. SUBSTRATES (0 m to 0.1 m):	**A. GROUND SURFACE (0 m to 0.5 m):**
• Litter/Light Fuel Consumed (0%/50%/100%)	Digital Terrain Model (in 0.25 m resolution)
• Duff (0%/50%/100%)	Structure (number of voxels in 0.25 m size)
• Medium Fuel, 7.62-20.32 cm (0%/20%40%/>60%)	No color
• Heavy Fuel, > 20.32 cm (0%/10%/25%/>40%)	No size classification
• Soil & Rock Cover/Color (0%/10%/40%/>80%)	
B. HERBS, LOW SHRUBS AND TREES (< 1 m):	**B. SHRUBS (0.5m to 1 m):**
• % Foliage Altered (blk-brn) (0%/30%80%/95%/100%)	Structure (number of voxels in 0.25 m size)
• Frequency % Living (100%/90%50%/< 20%/0%)	No identification of dead or alive
• Colonizers (0%/Low/Moderate/High-Low/Low to None)	No species classification
• Spp. Comp. - Rel. Abund. (0%/Little/Moderate/High change)	
C. TALL SHRUBS AND TREES (1 m to 5 m):	**C. SHRUBS AND UNDERSTORY (1 m to 5 m):**
• % Foliage Altered (0%/20%/60-90%/>95%)	Structure (number of voxels in 0.25 m size)
• Frequency % Living (100%/90%/30%/< 15%/< 1%)	No identification of dead or alive
• % Change in Cover (0%/15%/70%/90%/100%)	No species classification
• Spp. Comp. - Rel. Abund. (0%/Little/Moderate/High Change)	
D. INTERMEDIATE TREES	**D. OVERSTORY TREES (5m and above)**
(SUBCANOPY, POLE-SIZED TREES)	Structure (number of voxels in 0.25 m size)
• % Green (Unaltered) (100%/80%/40%/< 10%/0%)	No color
• % Black (Torch) (0%/5-20%/60%/> 85%/100%)	No identification of dead or alive
• % Brown (Scorch/Girdle) (0%/5-20%/40-80%/>80%)	No canopy class identification
• % Canopy Mortality (0%/15%/60%/80%/100%)	
• Char Height (None/1.5 m/2.8 m/> 5 m)	
E. BIG TREES	
(UPPER CANOPY, DOMINANT, CODOMNANT TREES)	
• % Green (Unaltered) (100%/95%/50%/< 10%/0%)	
• % Black (Torch) (0%/5-10%/50%/> 80%/100%)	
• % Brown (Scorch/Girdle) (0%/5-10%/30-70%/> 70%)	
• % Canopy Mortality (0%/10%/50%/70%/100%)	
• Char Height (None/1.8 m/4 m/> 7 m)	

The CBI is based on visual estimation and judgement to assess vegetation coverage at different height strata (Box 1). To simulate human vision and to derive values based more closely on the CBI protocol, all 3D data are converted to a spherical coordinate system to create a human vision-oriented image (Figures 2 and 3). Based on TLS sensor location, all XYZ 3D coordinates of voxels are converted to three variables in spherical coordinates: distance, horizontal angles (θ), and vertical angles (Φ). The degrees of horizontal and vertical angles are displayed in X and Y axis to make a human vision-oriented image shown in Figure 3. Then, the counted pixels on the image are compared with counted voxels to quantify the visual bias for the different height strata. The voxels displayed in orthogonal coordinates are paired with the satellite image displayed in plain view. To visualize the bias effect, the voxels in each height stratum in the orthogonal coordinates are counted in total, and the number of pixels (in spherical coordinates) visible on the human vision-oriented image in each height strata are also counted. To compare the voxelization method with our simulation of CBI field assessment, the counted total is normalized with the maximum number on a scale from 0 to 1, for each of the height strata, in each scan. A natural logarithmic equation is applied to obtain the non-linear relationship of the visual bias between x in spherical and y in orthogonal coordinates.

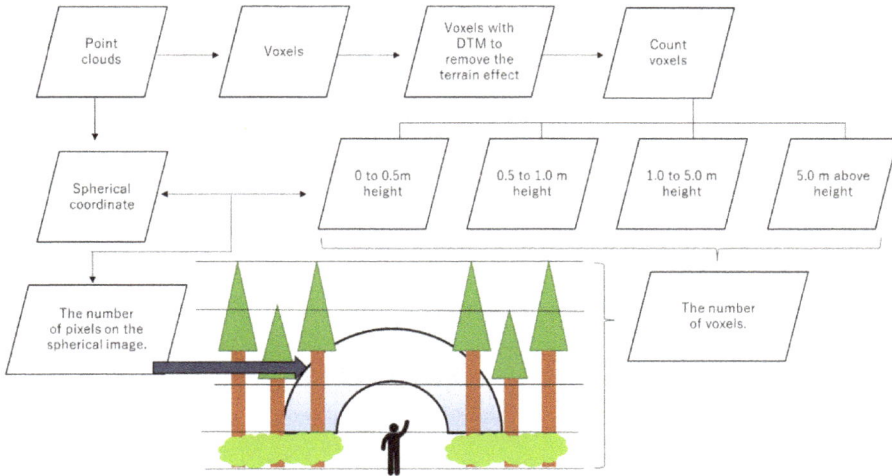

Figure 2. TLS data analysis procedure.

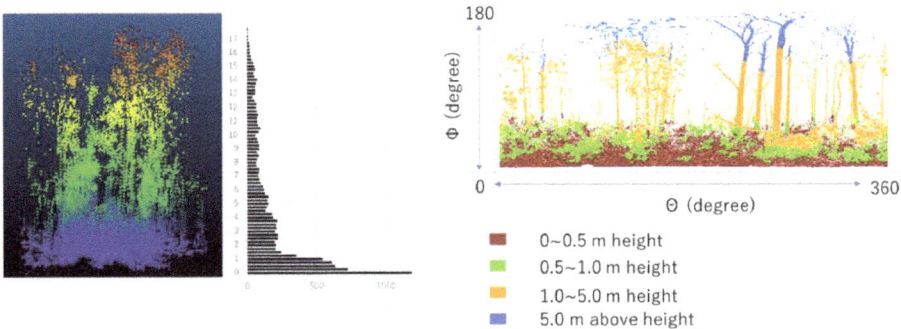

0~0.5 m height
0.5~1.0 m height
1.0~5.0 m height
5.0 m above height

Figure 3. Vertical voxel distribution and counting in orthogonal coordinate (left) and human vision-oriented image in spherical coordinates (right).

3. Results

There were 32 TLS scans in total. Eleven of these were in recently burned plots and 21 in unburned plots. There was a linear relationship between the total number of voxels within each scan in the unburned plots, and satellite derived dNBR (R^2 value of 0.54, p value < 0.05) and RdNBR values (R^2 value of 0.60, p value < 0.05) (Figure 4). In the burned plots, there was a non-linear correlation with R^2 values of 0.75 (p value > 0.05) and 0.73 (p value > 0.05) for dNRR and RdNBR respectively (Figure 4). Similar relationships of dNRR and RdNBR were compared to the number of voxels at each of the four height strata (Figures 5 and 6). There was a negative linear slope for unburned plots and a non-linear relationship for recently burned plots.

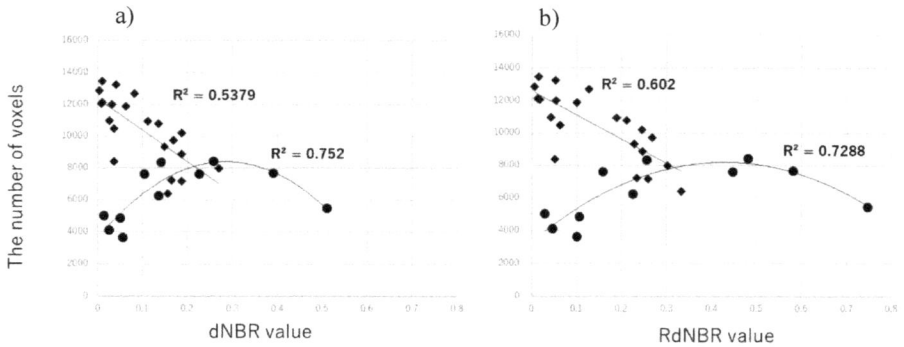

Figure 4. The relationship between the total number of voxels at each plot and (**a**) dNBR and (**b**) RdNBR values (marker ●: recently burned plots, ♦: unburned plots, the bold fonts of R^2 values are statistically significant for the linear case and not-significant for the non-linear case with a significant level of 0.05).

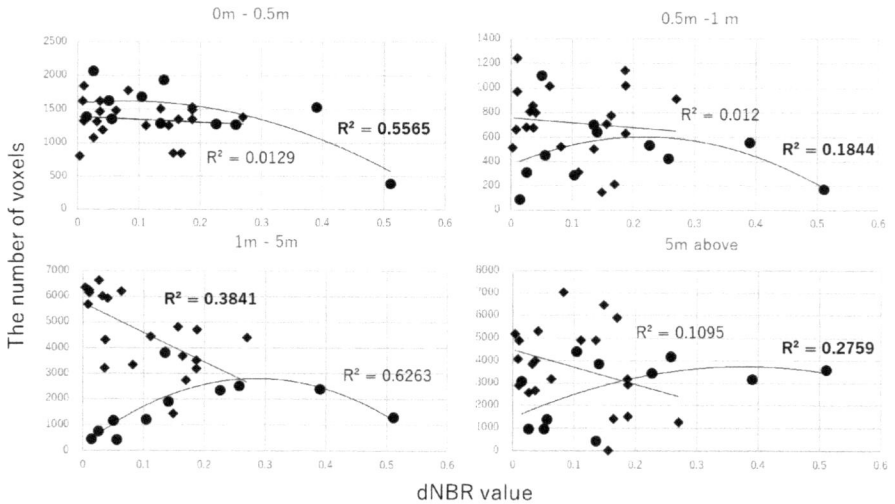

Figure 5. The relationship between dNBR and the number of voxels at different heights or strata (marker ●: recently burned plot, ♦: unburned plot, the bold fonts of R^2 values are statistically significant for the linear case and not-significant for the non-linear case with a significant level of 0.05).

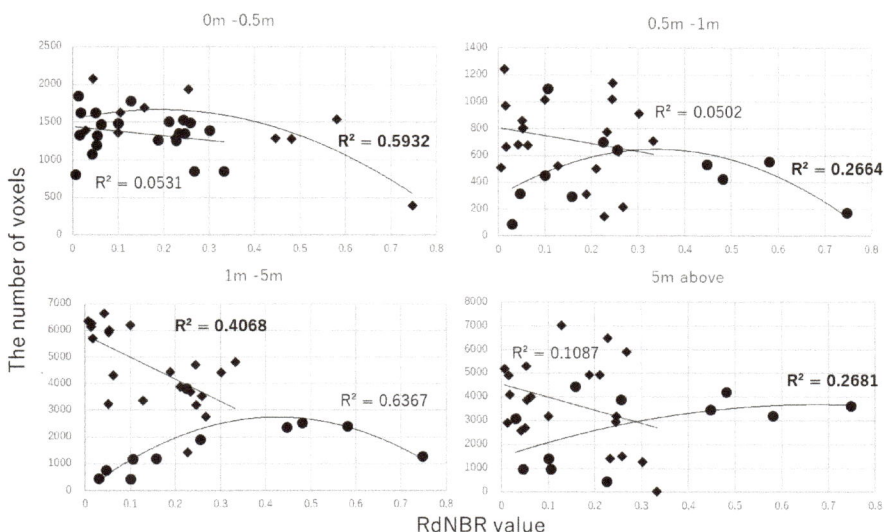

Figure 6. The relationship between RdNBR and the number of voxels at different heights or strata (marker ●: recently burned plot, ◆: unburned plot, the bold fonts of R^2 values are statistically significant for the linear case and not-significant for the non-linear case with a significant level of 0.05).

In the recently burned plots, the strongest relationship between voxel counts and dNBR and RdNBR values was in the 0 to 0.5 m height stratum with R^2 values of 0.56 for dNBR (p value > 0.05) and 0.59 for RdNBR (p value > 0.05). Additionally, for burned plots there were weak correlations in the 0.5 to 1 m and above 5 m height strata with R^2 values of 0.18 and 0.28 (p value > 0.05) for dNBR respectively and R^2 values of 0.27 and 0.27 (p value > 0.05) for RdNBR respectively. The unburned plots had the strongest relationship between voxel counts and dNBR and RdNBR values in the height stratum between 1 to 5 m, producing R^2 values of 0.38 for dNBR (p value < 0.05) and 0.41 for RdNBR (p value < 0.05) in Figures 5 and 6. The results from Figures 5 and 6 found specific height strata only related with the spectral change.

There was a non-linear relationship (visual bias) between the voxel approach and the pixel-based estimation in the three lower strata (0 to 0.5 m, 1 m to 1.5 m, and 1.5 m to 5 m) with a linear relationship in the upper (+5 m) strata (Figure 7). The bias was greatest in the lowest strata.

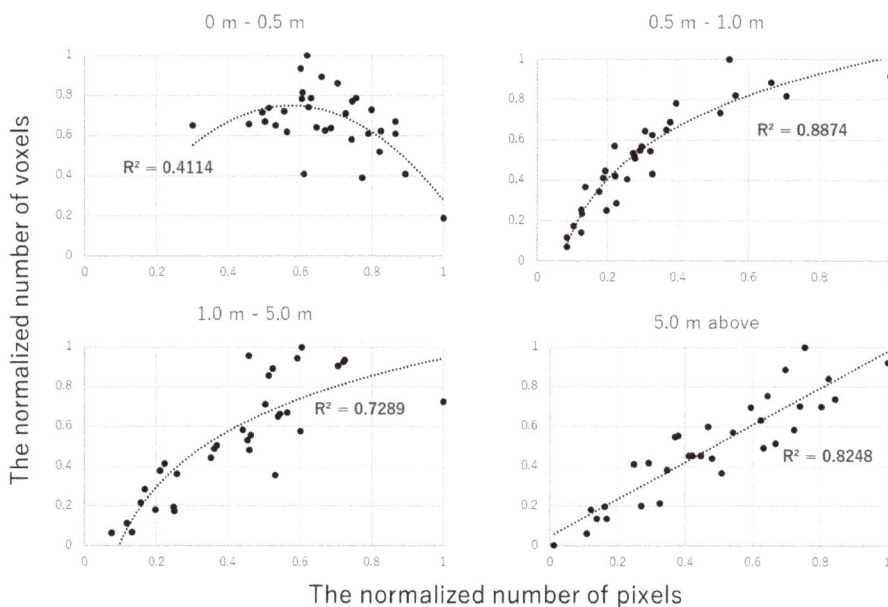

Figure 7. The relationship between the number of pixels in spherical coordinates (x) and the number of voxels in orthogonal coordinates (y) by different height strata (both axes are normalized by the maximum number and all R^2 values are statistically significant for the linear case and not-significant for the non-linear case with a significant level of 0.05).

4. Discussion

Comparing dNBR values with RdNBR and the total number of voxels per plot (Figure 4), the range of RdNBR was wider than the range of dNBR because of the effect of including pre-fire vegetation conditions in the equation. There was no significant difference between R^2 values presented in Figures 5 and 6.

For unburned plots, Figures 5 and 6 show that increased numbers of voxels in the height strata (1 to 5 m) were negatively correlated with dNBR and RdNBR, because more vegetation structure developed at the 1 to 5 m height stratum produced less spectral variability in dNBR and RdNBR. The post-fire NBR values approached the pre-fire NBR values due to the structural change at different height strata. For burned-plots, the negative correlation in the lowest (i.e., ground) strata indicated vegetation recovery of the strata. This process of using voxels improves not only the quality of ground observation data but also allows for better correlation with satellite images. Improved illumination values were collected from multi-temporal images through the image compositing process to compute dNBR and RdNBR, and higher precision 3D data were obtained to compare between burned and unburned plots.

The dNBR and RdNBR are relative measurements between pre and post fire NBR. If the site has experienced no fire, they are presumed to be zero. However, there were some changed values at unburned plots in Figures 5 and 6. The slight change was related to the structural difference of the height strata between 1 to 5 m. This shows an advantage to using high precision 3D data to detect small changes in that height stratum. This method is useful to describe ecological response to fire in burned plots and structural development in unburned plots. Through this process, the specific height strata related with the spectral change are identified by the stronger correlations.

For burned plots, the relationship between counted voxels and spectral indices was non-linear (Figures 4–6) and was related to the ecological responses of the site conditions. High spectral change (in *x* axis) with low voxel count (in *y* axis) characterizes the severely burned sites immediately after fire; middle spectral change with high voxel count means live vegetation remained after fire; and low spectral change with low voxel count means no vegetation recovery after fire.

The bias shown in Figure 7 helps to understand the non-linear relationship [8,14,38–41] between satellite image analysis such as dNBR and RdNBR and the ground visual assessment of CBI. Previous studies have shown that when simulated CBI was prepared from satellite images there was a linear relationship [42]. However, when the canopy fraction was included in the derived CBI [43], the relationship became non-linear. If the site has vegetation less than 5 m tall, then height strata need to be implemented in the severity measurement; the visual bias produces more deviation from the correlation between satellite and ground observation. The 3D structural data help to visualize the bias in observed vegetation coverage.

The utility of a 2D severity map from spectral indices such as dNBR and RdNBR is limited to assess severity. With 3D ground observations, the spectral change can be addressed by structural change. A good example of this application is fire refugia. The vegetation recovery takes different trajectoris with the initial structure remaining after fire. The severity assessment with 3D ground observation helps to characterize the difference.

Three years were required to detect an ecological response at this study site. The severity is measured by the most immediate fire effects as an immediate assessment and additional responses of initial severity as an extended assessment. The duration of vegetation recovery depends on the ecosystem and the climate of the site. The recovery rate is slow in boreal forest region and the recent fire burned with high intensity. With high intensity fire and slow vegetation recovery, the spectral change observed over burned plots is more related with ecological response. The three year interval was needed to detect the ecological response for fire severity. Monitoring ecological response after fire is an important application of TLS ground observation.

With regard to TLS ground sampling, one third of samples only came from burned plots (Figures 5 and 6). In this study site, the sampling locations were limited to places along the roads due to difficult accessibility. The number of burned plots facing roads were more limited than the number of unburned plots. That is why only one third of samples were from burned plots. The ground sampling strategy can be improved to capture more diverse fire severity.

A TLS limitation for collecting ground forest structure data includes occlusion of laser returns by forest obstacles within the scanning view angle [44,45]. In this study, a sampling approach is applied that isn't dependent on capturing occlusion free 3D data [16,17]. Our sampling strategy was to compare 3D data among plots by using a single scanning location. Multiple-scans per plot would be better to capture more complete structure measurements as occlusion is eliminated, but multiple scans invite more sampling variation with additional factors of scan positioning and angle being introduced. The single scanning strategy has been adopted to compare different vertical vegetation profiles and describe different structural conditions [46]. Calders and co-authors found that a single scanning was enough to describe various vertical plant profiles. This TLS study does not aim to provide absolute accuracy in tree measurements of the sites, but the results provides more accurate "relative differences" among plots without any subjective judgement, which could not be achieved by conventional visual estimates. This 3D ground truth is needed more in validating fine scale, post fire change when being estimated by high-resolution optical [47] and radar images [48].

5. Conclusions

Forest fire is measured by extent and severity of fire. The objective of this study was to propose a new ground observation technique using 3D data collected by Terrestrial Laser Scanner (TLS) in the field, and to determine its value for ground validation of spectral change on Landsat images. The spectral change between pre- and post-fire is compared with structural differences amongst the sites with different fire severities. To compare them, TLS generated 3D data was changed to voxels, and the number of voxels was counted and compared with dNBR and RdNBR in four different height strata. There was a negative linear relationship in the height strata between 1 m to 5 m for unburned plots and a non-linear relationship in the height strata between 0 to 0.5 m for burned plots. Shrub and understory development was detected in tall shrub strata for unburned plots, and vegetation recovery in the lowest height strata (0 to 0.5 m) was detected for burned-plots. Furthermore, there was a non-linear relationship between visual assessment of CBI and burn indices derived from satellite images. Fine resolution remote sensing imagery is commonly available and accessible through GEE, which will require a more accurate validation method from ground data collection. To make it efficient to collect data and to match the precision of high-resolution data, ground sampling using TLS to derive 3D data will play an important role in improving the correlation between satellite and field data. TLS is especially useful for quick ground assessments which are needed for forest fire applications.

Author Contributions: Conceptualization, A.K. and L.M.M.; methodology, A.K., D.T. and A.T.H.; software, A.K. and D.T.; validation, A.K.; formal analysis, A.K.; writing—original draft preparation, A.K.; writing—review and editing, J.L.B., D.T., L.M.M. and A.T.H.; funding acquisition, A.K.

Funding: This research was supported by the Environment Research and Technology Development Fund of the Ministry of the Environment, Japan, grant number 2RF-1501 and the UW Precision Forestry Cooperative.

Acknowledgments: The authors would like to acknowledge Canadian Park Services for providing the fire map and thank Garrett W. Meigs (Oregon State University, USA) for valuable advice on this research and reviews of this paper, Akira Osawa (Kyoto University) for proving research sites for this study, Yuichi Hayakawa (Hokkaido University, Japan) for proving the TLS sensor for this fieldwork, Mizuki Taga (Olympus Co. Ltd., Japan), and Masuto Ebina (Hokkaido Research Organization, Japan) for their field assistance of TLS data collection for this study.

Conflicts of Interest: The authors declare no conflict of interest.

References

1. Hudak, A.T.; Morgan, P.; Bobbitt, M.J.; Smith, A.M.S.; Lewis, S.A.; Letile, L.B.; Robichaud, P.R.; Clark, J.T.; McKinley, R.A. The Relationship of Multispectral Satellite Imagery to Immediate Fire Effects. *Fire Ecol.* **2007**, *3*, 64–90. [CrossRef]

2. Sugihara, N.G.; van Wagtendonk, J.W.; Shaffer, K.E.; Fites-Kaufman, J.; Thode, A.E. *Fire in Calfornia's Ecosystems*; University of Calfornia Press: Berkeley and Los Angeles, Calfornia, CA, USA, 2006.

3. Key, C.H.; Benson, N.C. Landscape assessment (LA): Sampling and analysis methods. FIREMON: Fire Effects Monitoring and Inventory System General Technical Report RMRS-GTR-164-CD. Fort Collins, CO: USDA Forest Service, Rocky Mountain Research Station. 2006.

4. García López, M.J.; Caselles, V. Mapping Burns and Natural Reforestation using Thematic Mapper Data. *Geocarto Intern.* **1991**, *6*, 31–37. [CrossRef]

5. Koutsias, N.; Karteris, M. Logistic regression modelling of multitemporal Thematic Mapper data for burned area mapping. **1998**, *1161*. [CrossRef]

6. Escuin, S.; Navarro, R.; Fernández, P. Fire severity assessment by using NBR (Normalized Burn Ratio) and NDVI (Normalized Difference Vegetation Index) derived from LANDSAT TM/ETM images. *Intern. J. Remote Sens.* **2008**, *29*, 1053–1073. [CrossRef]

7. Miller, J.D.; Thode, A.E. Quantifying burn severity in a heterogeneous landscape with a relative version of the delta Normalized Burn Ratio (dNBR). *Remote Sens. Environ.* **2007**, *109*, 66–80. [CrossRef]

8. Soverel, N.O.; Coops, N.C.; Perrakis, D.D.B.; Daniels, L.D.; Gergel, S.E. The transferability of a dNBR-derived model to predict burn severity across 10 wildland fires in western Canada. *Intern. J. Wildland Fire* **2011**, *20*, 518–531. [CrossRef]

9. Verbyla, D.L.; Kasischke, E.S.; Elizabeth, E.H. Seasonal and topographic effects on estimating fire severity from Landsat TM/ETM + data. *Intern. J. Wildland Fire* **2008**, *17*, 527–534. [CrossRef]

10. Allen, J.L.; Sorbel, B. Assessing the differenced Normalized Burn Ratio' s ability to map burn severity in the boreal forest and tundra ecosystems of Alaska ' s national parks. *Intern. J. Wildland Fire* **2008**, *17*, 463–475.

11. Kasischke, E.S.; Turetsky, M.R.; Ottmar, R.D.; French, N.H.F.; Hoy, E.E.; Kane, E.S. Evaluation of the composite burn index for assessing fire severity in Alaskan black spruce forests. *Intern. J. Wildland Fire* **2008**, *17*, 515–526. [CrossRef]

12. Murphy, K.A.; Reynolds, J.H.; Koltun, J.M. Evaluating the ability of the differenced Normalized Burn Ratio (dNBR) to predict ecologically significant burn severity in Alaskan boreal forests. *Intern. J. Wildland Fire* **2008**, *17*, 490–499. [CrossRef]

13. Lentile, L.B.; Holden, Z.A.; Smith, A.M.S.; Falkowski, M.J.; Hudak, A.T.; Morgan, P.; Lewis, S.A.; Gessler, P.E.; Benson, N.C. Remote sensing techniques to assess active fire and post fire effects: clarification of terminology. *Intern. J. Wildland Fire* **2006**, *15*, 319–345. [CrossRef]

14. Miller, J.D.; Knapp, E.E.; Key, C.H.; Skinner, C.N.; Isbell, C.J.; Creasy, R.M.; Sherlock, J.W. Calibration and validation of the relative differenced Normalized Burn Ratio (RdNBR) to three measures of fire severity in the Sierra Nevada and Klamath Mountains, California, USA. *Remote Sens. Environ.* **2009**, *113*, 645–656. [CrossRef]

15. Dixon, G.E. *Essential FVS: A User's Guide to the Forest Vegetation Simulator*; USDA-Forest Service, Forest Management Service Center: Fort Collins, CO, USA, 2002.

16. Hackenberg, J.; Spiecker, H.; Calders, K.; Disney, M.; Raumonen, P.; Observation, E.; Group, O.; Road, H.; Street, G.; Liang, X.; et al. SimpleTree—An Efficient Open Source Tool to Build Tree Models from TLS Clouds. *Forests* **2015**, *92*, 4245–4294. [CrossRef]

17. Raumonen, P.; Kaasalainen, M.; Akerblom, M.; Kaasalainen, S.; Kaartinen, H.; Vastaranta, M.; Holopainen, M.; Disney, M.; Lewis, P. Fast Automatic Precision Tree Models from Terrestrial Laser Scanner Data. *Remote Sens.* **2013**, *5*, 491–520. [CrossRef]

18. Seidel, D.; Beyer, F.; Hertel, D.; Fleck, S.; Leuschner, C. 3D-laser scanning: A non-destructive method for studying above- ground biomass and growth of juvenile trees. *Agric. For. Meteorol.* **2011**, *151*, 1305–1311. [CrossRef]

19. Zheng, G.; Moskal, L.M. Computational-Geometry-Based Retrieval of Effective Leaf Area Index Using Terrestrial Laser Scanning. *IEEE Trans. Geosci. Remote Sens.* **2012**, *50*, 3958–3969. [CrossRef]

20. Seidel, D.; Ammer, C.; Puettmann, K. Describing forest canopy gaps efficiently, accurately, and objectively: New prospects through the use of terrestrial laser scanning. *Agric. For. Meteorol.* **2015**, *213*, 23–32. [CrossRef]

21. Paynter, I.; Saenz, E.; Genest, D.; Peri, F.; Erb, A.; Li, Z.; Wiggin, K.; Muir, J.; Raumonen, P.; Schaaf, E.S.; et al. Observing ecosystems with lightweight, rapid-scanning terrestrial lidar scanners. *Remote Sens. Ecol. Conserv.* **2016**. [CrossRef]

22. Hopkinson, C.; Chasmer, L.; Young-Pow, C.; Treitz, P. Assessing forest metrics with a ground-based scanning lidar. *Can. J. For. Res.* **2004**, *34*, 573–583. [CrossRef]

23. Stovall, A.E.L.; Anderson-Teixeira, K.J.; Shugart, H.H. Assessing terrestrial laser scanning for developing non-destructive biomass allometry. *For. Ecol. Manag.* **2018**, *427*, 217–229. [CrossRef]

24. Lin, Y.; Herold, M. Tree species classification based on explicit tree structure feature parameters derived from static terrestrial laser scanning data. *Agric. For. Meteorol.* **2016**, *216*, 105–114. [CrossRef]

25. Lim, Y.; Hyyppä, J.; Kukko, A.; Jaakkola, A.; Kaartinen, H. Tree Height Growth Measurement with Single-Scan Airborne, Static Terrestrial and Mobile Laser Scanning. *Sensors* **2012**, *12*, 12798–12813. [CrossRef]

26. Calders, K.; Schenkels, T.; Bartholomeus, H.; Armston, J.; Verbesselt, J.; Herold, M. Monitoring spring phenology with high temporal resolution terrestrial LiDAR measurements. *Agric. For. Meteorol.* **2015**, *203*, 158–168. [CrossRef]

27. Metz, J.; Seidel, D.; Schall, P.; Scheffer, D.; Schulze, E.-D.; Ammer, C. Crown modeling by terrestrial laser scanning as an approach to assess the effect of aboveground intra- and interspecific competition on tree growth. *For. Ecol. Manag.* **2013**, *310*, 275–288. [CrossRef]

28. Srinivasan, S.; Popescu, S.C.; Eriksson, M.; Sheridan, R.D.; Ku, N.-W. Multi-temporal terrestrial laser scanning for modeling tree biomass change. *For. Ecol. Manag.* **2014**, *318*, 304–317. [CrossRef]

29. Gupta, V.; Reinke, K.J.; Jones, S.D.; Wallace, L.; Holden, L. Assessing Metrics for Estimating Fire Induced Change in the Forest Understorey Structure Using Terrestrial Laser Scanning. *Remote Sens.* **2015**, 8180–8201. [CrossRef]

30. Campos-Ruiz, R.; Parisien, M.-A.; Flannigan, M.D. Temporal patterns of wildfire activity in areas of contrasting human influence in the Canadian boreal forest. *Forests* **2018**, *9*. [CrossRef]

31. Kochtubajda, B.; Flannigan, M.D.; Gyakum, J.R.; Stewart, R.E.; Logan, K.A.; Nguyen, T.V. Lightning and fires in the Northwest Territories and responses to future climate change. *Arctic* **2006**, *59*, 211–221. [CrossRef]

32. Gillett, N.P.; Weaver, A.J.; Zwiers, F.W.; Flannigan, M.D. Detecting the effect of climate change on Canadian forest fires. *Geophys. Res. Lett.* **2004**, *31*. [CrossRef]

33. Stocks, B.J.; Mason, J.A.; Todd, J.B.; Bosch, E.M.; Wotton, B.M.; Amiro, B.D.; Flannigan, M.D.; Hirsch, K.G.; Logan, K.A.; Martell, D.L.; et al. Large forest fires in Canada, 1959–1997. *J. Geophys. Res.* **2002**, *108*. [CrossRef]

34. Osawa, A. Inverse relationship of crown fractal dimension to self-thinning exponent of treepopulations: A hypothesis. *Can. J. For. Res.* **1995**, *25*, 1608–1617. [CrossRef]

35. Osawa, A.; Kurachi, N. Spatial leaf distribution and self-thinning exponent of Pinus banksiana and Populus tremuloides. *Trees* **2004**, *18*, 327–338. [CrossRef]

36. Gorelick, N.; Hancher, M.; Dixon, M.; Ilyushchenko, S.; Thau, D.; Moore, R. Google Earth Engine: Planetary-scale geospatial analysis for everyone. *Remote Sens. Environ.* **2017**, *202*, 18–27. [CrossRef]

37. Shao, Y.-C.; Chen, L.-C. Automated Searching of Ground Points from Airborne Lidar Data Using a Climbing and Sliding Method. *Photogramm. Eng. Remote Sens.* **2008**, *74*, 625–635. [CrossRef]

38. Hall, R.J.; Freeburn, J.T.; de Groot, W.J.; Pritchard, J.M.; Lynham, T.J.; Landry, R. Remote sensing of burn severity: Experience from western Canada boreal fires. *Intern. J. Wildland Fire* **2008**, *17*, 476–489. [CrossRef]

39. Key, C.H. Ecological and Sampling Constraints on Defining Landscape Fire Severity. *Fire Ecol.* **2006**, *2*, 34–59. [CrossRef]

40. Van Wagtendonk, J.W.; Root, R.R.; Key, C.H. Comparison of AVIRIS and Landsat ETM+ detection capabilities for burn severity. *Remote Sens. Environ.* **2004**, *92*, 397–408. [CrossRef]

41. Zhu, Z.; Key, C.H.; Ohlen, D.; Benson, N. Evaluate Sensitivities of Burn-Severity Mapping Algorithms for Different Ecosystems and Fire Histories in the United States. Final Report to the Joint Fire Science Program. JFSP Project No. 01-1-4-12. Sioux Falls, SD: USGS, National Center for Earth Resources Observation and Science. 2006.

42. Chuvieco, E.; Riaño, D.; Danson, F.M.; Martin, P. Use of a radiative transfer model to simulate the postfire spectral response to burn severity. *J. Geophys. Res. Biogeosci.* **2006**, *111*, 1–15. [CrossRef]

43. De Santis, A.; Chuvieco, E. GeoCBI: A modified version of the Composite Burn Index for the initial assessment of the short-term burn severity from remotely sensed data. *Remote Sens. Environ.* **2009**, *113*, 554–562. [CrossRef]

44. Zande, D.V.D.; Hoet, W.; Jonckheere, I.; Aardt, J.V.; Coppin, P. Influence of measurement set-up of ground-based LiDAR for derivation of tree structure. *Agric. For. Meteorol.* **2006**, *141*, 147–160. [CrossRef]

45. Zande, D.V.D.; Jonckheere, I.; Stuckens, J.; Verstraeten, W.W.; Coppin, P. Sampling design of ground-based lidar measurements of forest canopy structure and its effect on shadowing. *Can. J. Remote Sens.* **2009**, *34*, 526–538. [CrossRef]

46. Calders, K.; Armston, J.; Newnham, G.; Herold, M.; Goodwin, N. Implications of sensor configuration and topography on vertical plant profiles derived from terrestrial LiDAR. *Agric. For. Meteorol.* **2014**, *194*, 104–117. [CrossRef]

47. Meng, R.; Wu, J.; Schwager, K.L.; Zhao, F.; Dennison, P.E.; Cook, B.D.; Brewster, K.; Green, T.M.; Serbin, S.P. Using high spatial resolution satellite imagery to map forest burn severity across spatial scales in a Pine Barrens ecosystem. *Remote Sens. Environ.* **2017**, *191*, 95–109. [CrossRef]

48. Tanase, M.A.; Kennedy, R.; Aponte, C. Radar Burn Ratio for fire severity estimation at canopy level: An example for temperate forests. *Remote Sens. Environ.* **2015**, *170*, 14–31. [CrossRef]

forests

MDPI

Article

Mapping Maximum Tree Height of the Great Khingan Mountain, Inner Mongolia Using the Allometric Scaling and Resource Limitations Model

Yao Zhang [1], Yuli Shi [1,*], Sungho Choi [2], Xiliang Ni [3] and Ranga B. Myneni [2]

[1] School of Remote Sensing and Geomatics Engineering, Nanjing University of Information Science and Technology, Nanjing 210044, China; yao_zhang_ll@outlook.com
[2] Department of Earth and Environment, Boston University, Boston, MA 02215, USA; schoi@bu.edu (S.C.); rmyneni@bu.edu (R.B.M.)
[3] State Key Laboratory of Remote Sensing Science, Institute of Remote Sensing and Digital Earth, Chinese Academy of Science, Beijing 100101, China; nixl@irsa.ac.cn
* Correspondence: ylshi.nuist@gmail.com; Tel.: +1-585-184-8199

Received: 19 March 2019; Accepted: 28 April 2019; Published: 30 April 2019

check for
updates

Abstract: Maximum tree height is an important indicator of forest vegetation in understanding the properties of plant communities. In this paper, we estimated regional maximum tree heights across the forest of the Great Khingan Mountain in Inner Mongolia with the allometric scaling and resource limitations model. The model integrates metabolic scaling theory and the water–energy balance equation (Penman–Monteith equation) to predict maximum tree height constrained by local resource availability. Monthly climate data, including precipitation, wind speed, vapor pressure, air temperature, and solar radiation are inputs of this model. Ground measurements, such as tree heights, diameters at breast height, and crown heights, have been used to compute the parameters of the model. In addition, Geoscience Laser Altimeter System (GLAS) data is used to verify the results of model prediction. We found that the prediction of regional maximum tree heights is highly correlated with the GLAS tree heights ($R^2 = 0.64$, RMSE = 2.87 m, MPSE = 12.45%). All trees are between 10 to 40 m in height, and trees in the north are taller than those in the south of the region of research. Furthermore, we analyzed the sensitivity of the input variables and found the model predictions are most sensitive to air temperature and vapor pressure.

Keywords: maximum forest heights; metabolic scale theory; allometric scaling and resource limitation model

1. Introduction

Forests, as a crucial part of terrestrial vegetation, play a central role in regulating the carbon and water cycles [1–4]. Moreover, height is an important indicator of various forest features, such as biological productivity, mortality rates, canopy density, and energy exchange [5–9].

Several articles have reported nonphysical or nonphysiological approaches to generate spatially continuous maps of forest heights by combining remote sensing data and in situ measurements. It is possible to estimate tree height with optical data and altimeter data from terrestrial, airborne [10,11], and spaceborne LiDAR [12–16]. Airborne LiDAR and stereo-photogrammetry data can effectively reflect the vertical structure of forests, but its application is constrained to small regional scale due to the expensive costs [17]. While the spaceborne LiDAR can provide global elevation information, the sampling density is insufficient for the complete monitoring of equatorial and midlatitude forests [18]. In addition, the underlying physical and biological principles of forest growth are often ignored in those approaches, and such neglect may lead to nonmechanistic shifts in the modelled outputs that are easily affected by the quality and quantity of training data [19].

Recent studies have applied spatial statistics and biophysical theories to establish biophysical models [20–23]. The model can give spatially continuous canopy heights of forests at large scale with the sparse observations and geospatial predictors like climatic variables and topography [20]. Climatic variables are good candidates for predictors of such models based on an assumption that climate regulates overall plant growth [24–26]. The model we used here, called allometric scaling and resource limitations (ASRL), is a biophysical model. The ASRL model integrates metabolic scaling theory (MST) for plants [27] and the water–energy balance equation [28] to predict potential maximum tree heights [29,30]. In ASRL model, the biophysical principles provide a generalized mechanistic understanding of relationships between tree size and geospatial parameters, including topography and climatic variables [29]. This model can be used for monitoring forests at large scales.

However, the original model is not suitable for some study areas due to differences in forest growth status, such as canopy density, stand age, and stand density [20,31]. In order to solve this problem, the ASRL model was improved in this paper to be highly consistent with the forest growth status in the study area. The improved ASRL model was used to map continuous maximum forest canopy heights of the Greater Khingan Mountain in Inner Mongolia with actual measurements, climatic data, and remote sensing data.

2. Data

The study area is situated in the Great Khingan Mountain, located within cold temperate continental monsoon climate zone of northeast Inner Mongolia, China (119°36′–125°24′ E, 47°03′–53°20′ N). It is hot and humid in summer, but cold and dry in winter. The annual average temperature is about −3.5 °C, while extreme low temperature can reach −50 °C. The annual mean precipitation in the study area is approximately 300–450 mm. The main forest in the study area is a mix of *Larix gmelinii* and White birch, which is formed by White birch's invasion after the destruction of the native *Larix gmelinii* forest. The forest covers approximately 8.17 million ha, with an elevation range 250–1745 m above sea level.

Field measurement data were derived from the ground survey data in Genhe city in August 2013 and 2016. Ninety plots were established and measured, including 19 square plots (45 × 45 m, or 30 × 30 m) and 71 circular plots (radius = 10, or 15 m). Figure 1 presents the distribution of these plots. The centers of each plot were located using Trimble GeoHX6000 Handheld GPS (Trimble, Sunnyvale, CA, USA) with an accuracy of approximately 2–3 m. Within each plot, diameter at breast height (DBH) of all live trees were measured using a diameter tape but only DBH over 5 cm were recorded. Trupulse TM2000 was used to measure tree height and height to crown base for each stand tree. Crown widths were approximated to the average of two values measured along two perpendicular directions from the location of the tree top. In order to avoid double counting of trees, latest record was used if a tree was measured more than once.

For input climate data, including monthly precipitation, wind speed, vapor pressure, air temperature and solar radiation, we used the WorldClim Version 2.0 (Sustainable Intensification Innovation Lab, Kansas State University, Manhattan, KS, USA) dataset averaged over multiple years from 1970 to 2000 at a 1-km spatial resolution (http://worldclim.org/version2). The input elevation data were derived from ASTER Global Digital Elevation Map (GDEM) V2 at a 30-m spatial resolution. All input gridded data were resampled and reprojected at a 1-km spatial resolution with a Lambert Conformal Conic map project to generate the continuous map of tree heights.

Two types of Moderate Resolution Imaging Spectroradiometer (MODIS) products were used as ancillary data in this study. Vegetation classification based on IGBP [32] derived from MODIS land cover type product (MCD12Q1) at a 500-m spatial resolution was used to define forest area (Figure 2a). Another ancillary data named MODIS Vegetation Continuous Filed (VCF) at 250-m spatial resolution was used to identify forest land with percent tree cover over 40 (Figure 2b). The ancillary data was at the same spatial resolution and had the same projection as the input data.

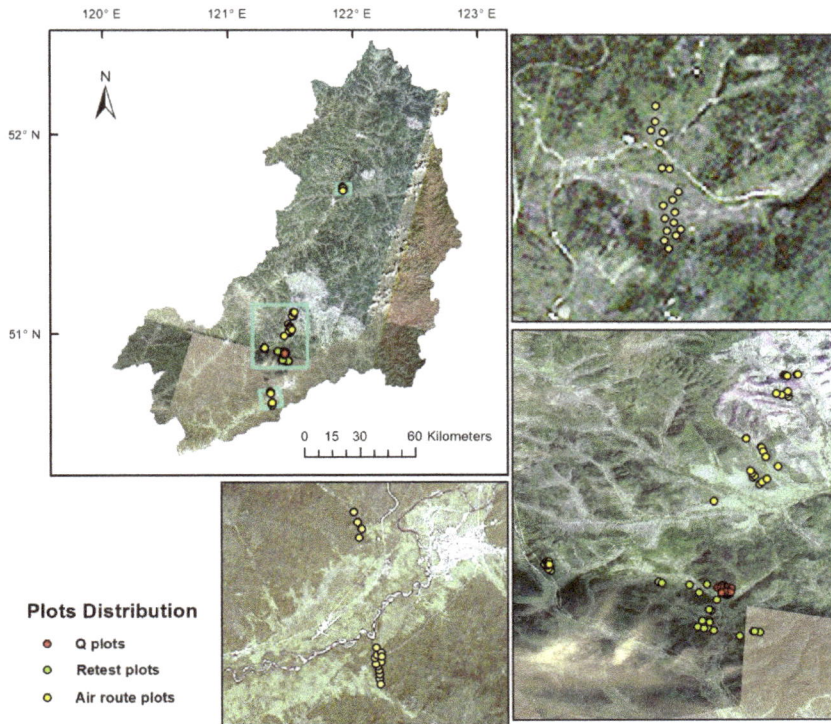

Figure 1. Distribution of plots in Genhe city.

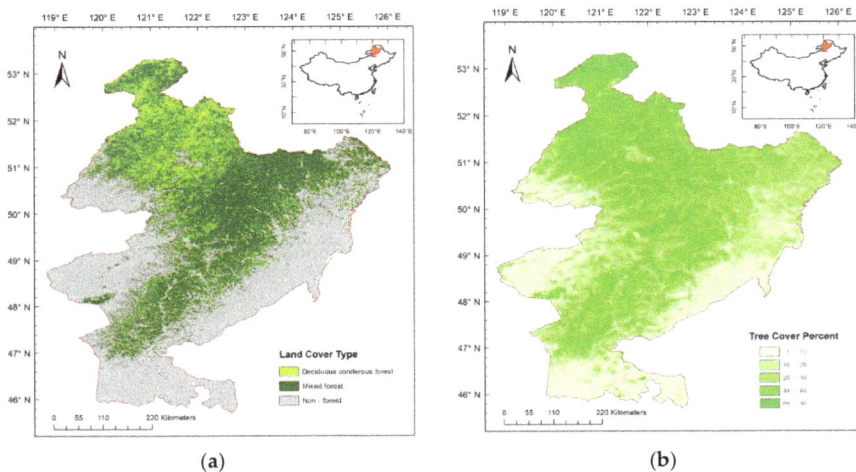

Figure 2. (**a**) Distribution of five surface cover types of the Greater Khingan Mountain in Inner Mongolia in 2013; (**b**) Vegetation coverage rate of the Greater Khingan Mountain in Inner Mongolia in 2013.

Global Surface Altimetry Data (GLA14 product) from 2003 to 2005 was used to extract maximum tree heights to verify predictions of the ASRL model. The distribution of GLAS footprints is in Figure 3. According to Ni's [23] research, when slope is smaller than 10, GLA14 product performs highest accuracy in maximum tree height's extraction. The best equation to estimate forest heights is:

$$H = (W_{SB} - W_{GP}) - d * \frac{\tan \theta}{2} \tag{1}$$

where W_{SB} represent the signal beginning and W_{GP} is the ground peak of GLAS full-waveform. While d is spot size and θ is topographic slope.

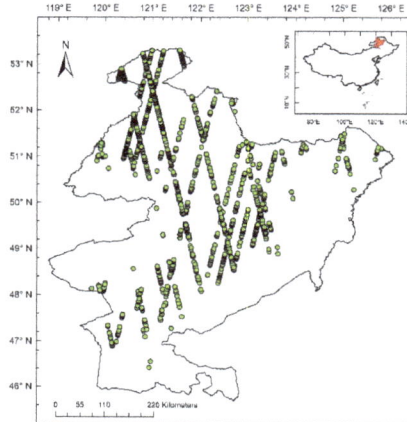

Figure 3. Distribution of GLAS footprints in the Great Khingan Mountain, Inner Mongolia.

3. Methods

3.1. The ASRL Model Framework

Biologists have found that the size and structure of living organisms have a great influence on their physiological processes [33,34]. In order to meet the needs of physiological processes, there is a stable proportional coefficient among the internal structure of the organism that accompanies its growth. The MST assumes that the plant metabolic rate B scales with the size of the whole plant, including volume V and mass M as: $B \propto V^\theta \propto M^\theta$ [35], and the parameter θ is close to 3/4. Kempes C.P. et al. [20] proposed ASRL tree height model which combines the metabolic scaling theory and energy balance equation. The ASRL model assumes that: (1) the tree can extract the resources from the environment which are needed for growth; (2) the ability of absorbing resources depends on the size of the tree; and (3) the resources that the environment can supply limit the growth of the tree. In the model, this is expressed by inequalities of three flow rates: $Q_0 \leq Q_e \leq Q_p$. The evaporation flow rate (Q_e) of a tree must satisfy its minimum metabolic flow rate (Q_0) but not exceed the potential rate of water inflow (Q_p) that the external environment can provide. These water flow rates are affected by both tree size and local environment supply. Based on the scale growth theory, we can use tree height to represent other characteristics of the tree, and the water and energy in the environment can be calculated by climatic predictors (such as temperature, pressure, water vapor pressure, solar radiation and precipitation). The basic framework of the model is shown in Figure 4.

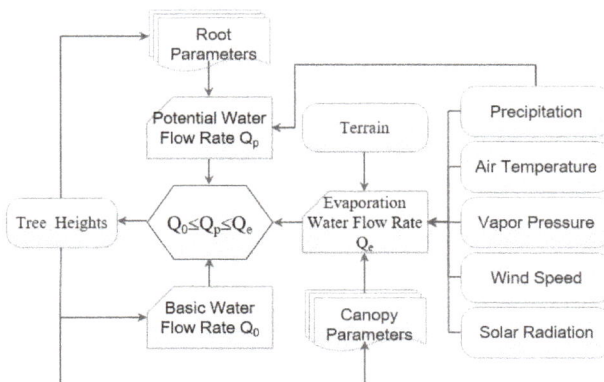

Figure 4. The basic framework of the allometric scaling and resource limitation (ASRL) model.

The basic water flow rate Q_0 is expressed as the equation of the tree height h:

$$Q_0 = \sum_{}^{12months} \beta_1 h^{\eta_1} \tag{2}$$

where β_1 and η_1 are the constant and exponent for basal metabolism. The potential water flow rate Q_p is based on tree height h, elevation, and precipitation:

$$Q_p = \sum_{}^{12months} \gamma \left(2\pi r_{root}^2\right) \Psi P_{inc} \tag{3}$$

The root absorption efficiency γ is related to local soil properties and terrain, and the $2\pi r_{root}^2$ is hemispheric root surface area [36,37]. The normalized terrain index Ψ is calculated from the elevation data, and P_{inc} is the input precipitation data. Evaporation water flow rate is given as a function of tree height h and climatic variables, including wind speed, solar radiation, temperature, precipitation, and vapor pressure:

$$Q_e = a_L v_{water} \sum_{}^{12months} E_{flux} \tag{4}$$

The effective tree area a_L is calculated from the single leaf area s_{leaf} and the branching architecture [20]. The molar volume of water v_{water} can be calculated from the molar mass of water and the water density. The evaporative molar flux E_{flux} is related to climatic factors such as temperature, water vapor pressure, and wind speed.

3.2. Improvements in the ASRL Model

Previous studies have found reasons for deviations from basic MST, including tree species, plant interaction, self-competition, and forest age [38,39]. The correlation established in the original model is difficult to reproduce in different research areas or times. According to Choi's [29] research, this paper makes the following improvements to the ASRL model to adapt to the research area. Key parameters in the ASRL model are presented in Table 1.

First, the growth coefficient of tree height and DBH in the MST model ($h \propto r_{stem}^{\phi}$, $\phi \approx 2/3$) is replaced with a statistic value of 0.7153. The measured tree height and DBH data is used to reconstruct the forest allometric scaling equation of the Greater Khingan Mountain in Inner Mongolia and replace the theoretical value of ϕ in the MST model. It reflects the differences in forest metabolism and metabolic variability in different regions [38,39].

Second, we replaced the scale factor of tree height h and crown height h_{cro} in the MST model ($h_{cro} \approx 0.79h$) with a statistic value of 0.47. Trees need to change their crown geometries and metabolic properties for the interplant interactions and self-competition [40,41]. The relationship between tree height and crown height in MST is unreliable, especially in the virgin forests of complex growth. The measured tree height and crown height data is used to reconstruct the forest allometric scaling equation of the Greater Khingan Mountain in Inner Mongolia and replace the theoretical value of 0.79 in the MST model.

Third, a dimensionless normalized topographic index Ψ is introduced to reflect local terrain features. Generally, the flow of water always flows from high to low, and the terrain will inevitably affect the collection of water flow. In this paper, we introduced a dimensionless topographic index to simulate the situation:

$$\Psi = \ln[CA/\tan(slp)]/ln[CA_0/\tan(slp_0)] \tag{5}$$

where CA is catchment area and slp is terrain slope. Assuming that the catchment area at hill top: CA_0 is 1, and slope at flat: slp_0 is e^{-10}. The topographic index of each pixel is calculated with DEM data, indicating the collection of precipitation due to effect of slopes.

Fourth, the canopy is treated as a huge leaf, and the energy exchange of the whole-plant is calculated based on the PM equation [28]. The soil heat flux G is also added into the energy balance:

$$R_{abs} = L + G + H + \lambda E_{flux} \tag{6}$$

where, the R_{abs} is absorbed solar radiation, L is thermal heat, and H is sensible heat.

Fifth, based on the measured tree height data, β_1, γ, and s$_{leaf}$ are optimized. In the ASRL model, β_1 is metabolic coefficient of a tree, and its theoretical value is 0.017, which is determined by the biological mechanism of a tree. γ is water absorption rate of roots with a theoretical value of 0.5. The value of water absorption rate may change in some soil types and environments. s$_{leaf}$ is the area of a single-leaf with a theoretical value of 0.001. Accompanying the tree's growth, the single leaf area will gradually change. These three parameters can't be obtained by direct measurement or calculation, but are important to the model: the basic water flow rate Q_0 is determined by β_1, while the value of γ can affect the potential water flow rate Q_p, and the size of s$_{leaf}$ can determine the water and energy metabolism rate of the whole tree. In order to obtain these three parameters, a nonlinear multivariate optimization equation is constructed:

$$D\{\beta_1, \gamma, S_{leaf}\} = \sum_n \left\{ \left[h_{obs} - h_c(\beta_1, \gamma, S_{leaf}) \right]^2 \right\} \tag{7}$$

where, h_{obs} is measured tree height and h_c is the modeled tree height. By iteration, when the D value is the minimum, the parameters are considered optimal.

Table 1. Key parameters in the ASRL model compared with previous studies.

Parameters	Description	Initial values	Optimized Values
Φ	Exponent for tree height and stem radius allometry	2/3	0.7153
β_1	Normalization constant for the basal metabolism	0.0177	0.005
γ	Water absorption efficiency	0.5	0.31
Ψ	Topographic index	\	Calculated by slope and catchment area
β_3	Crown ratio	0.79	0.47
s$_{leaf}$	Area of single leaf	0.001	0.0004

4. Results

Based on the improved ASRL tree height model, we generated the map of maximum tree heights of the Great Khingan Mountain in Inner Mongolia (Figure 5a). Tree heights in the research area are not more than 40 m. Trees in the north are taller than those in the south. Modelling tree heights are verified with the GLAS tree heights in the research area, and the results are shown in Figure 5b–d. The maximum tree height in ASRL predictions has a statistically significant linear relationship with the GLAS height (R^2 = 0.64, RMSE = 2.87 m, PMSE = 12.45%).

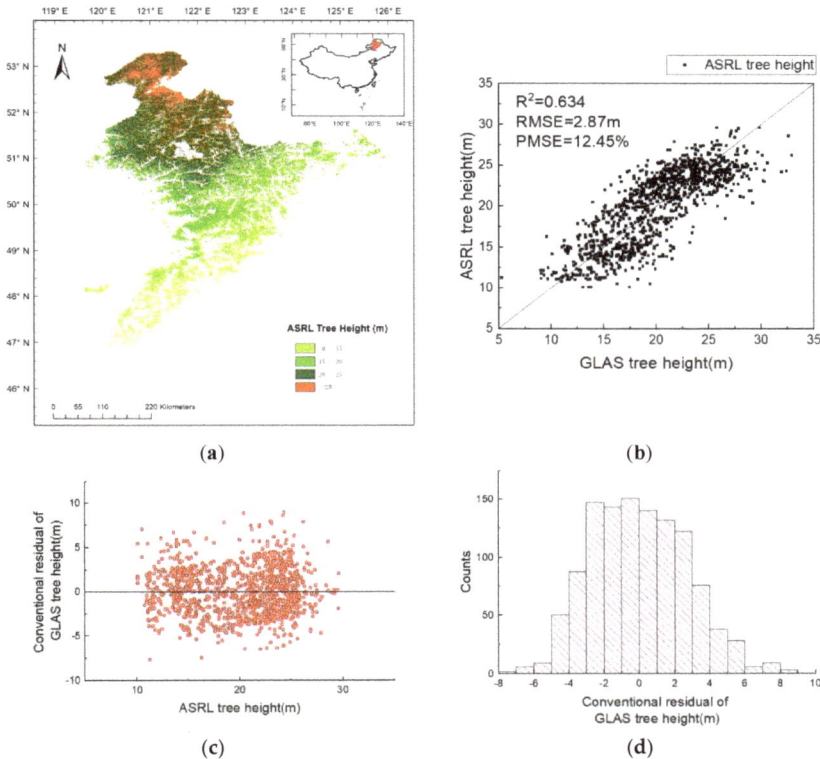

Figure 5. Inversion and verification results of the ASRL tree height model. (**a**) The distribution map of the maximum tree heights of the Great Khingan Mountain in Inner Mongolia based on the predictions of improved ASRL model (unit: m). (**b–d**) Three kinds of verification results: (**b**) The linear fitting of ASRL tree heights and GLAS tree heights (R^2 = 0.634, RMSE = 2.87m, PMSE = 12.45%); (**c**) The residual distribution of ASRL tree heights to GLAS tree heights, and (**d**) the counts of (**c**).

5. Discussions

5.1. Model Improvement

Kempes' model is based on metabolic scaling theory and resource constraint theory, and theoretical values of model parameters are given and applied to tree height calculations. In real-world applications of this model, these parameters have to be replaced and optimized to make the tree height model highly consistent with the forest growth status in the study area. The optimization of the model includes three items: parameter replacement, parameter optimization, and introduction of new parameters. The point with coordinates 121.554° E and 53.291° N is selected as the verification point to verify the results of each optimization by controlling variables. The measured tree height of the verification point is 24.6 m. Climatic data of the verification point are imported into the model before and after optimization, and the inversion results are compared and analyzed. Therefore, this paper constructs the ASRL model in four cases: no parameters replacement (NPR), no parameters optimization (NPO), no topographic index (NTI), and the optimized model (OM).

In the original ASRL tree height model, the tree height h and DBH r_{stem}. meets the following rule: $h \propto r_{stem}^{\phi}, \phi \approx 2/3$. Enquist et al. and Kempes et al. found that crown height h_{crow} and tree height h are required to be: $h_{cro} \approx \beta_3 h, \beta_3 = 0.79$. In order to improve the fit degree of the model to the research area, this paper utilizes the field data of tree heights, DBH and crown height in the Genhe city to establish the growth relationship between DBH, crown height and tree height, respectively. The results are shown in Figure 6. According to the measured data, parameter ϕ was 0.715, and the growth coefficient of crown height and tree height is $h_{cro} \approx 0.47h$.

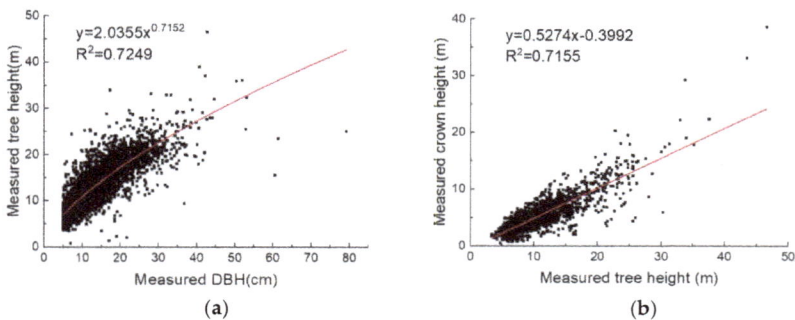

Figure 6. Modeling results of measured data. (**a**) The linear fitting result of measured tree height and diameter at breast height (DBH) ($R^2 = 0.7249$, $\Phi = 0.7249$). (**b**) The linear fitting result of measured tree height and crown height ($R^2 = 0.7155$, $\beta_3 = 0.47$).

A cost function D solved by the constrained nonlinear multivariable optimization is used to optimize the three parameters: β_1, γ, and S_{leaf}. The initial ASRL parameters were $\beta_1 = 0.01$, $\gamma = 0.5$ and $s_{leaf} = 0.001$ [7,20]. Inputting maximum tree height of each measured field as a sample, we minimized the cost function D by calibrating all three parameters within ranges ($0.005 < \beta_1 < 0.02$, $0.01 < \gamma < 1$ and $0.0001 < s_{leaf} < 0.01$). Finally, the optimal parameters obtained in this paper were $\beta_1 = 0.005, \gamma = 0.31, S_{leaf} = 0.0004$.

With no parameter changes, including NPR (Figure 7b) and NPO (Figure 7c), the ASRL predictions at verification point are smaller than actual measurement. Comparing the curves of three kinds of water flow rates, the basic water flow rate and the potential water flow rate are not affected, but the actual evaporation water flow rate is significantly increased, which leads the intersection point of Q_p and Q_e to shift left and the predicted tree height to be smaller. The prediction of model without the normalized topographic index Ψ is 33.2 m. As Figure 7d shows, the potential water flow rate is clearly increased, which leads the intersection point of Q_p and Q_e to shift right and the inversion result to be higher. Comparing with the result of the optimized model (Figure 7a), which is 22.7 m, we found: (1) parameter adjustment can make the result of evaporation water flow rate more reasonable and it effectively avoids underestimation of high trees; (2) the introduction of normalized topography index can reduce the sink flow in the high-terrain area and increase the sink flow in the low-terrain area, so that the convergence of precipitation on the surface in the model is consistent with the actual situation and the prediction accuracy of tree height has been improved. In addition, the curves of the optimized model show that maximum potential tree heights are mainly limited by water supply, meaning the verification point is a water-limited environment.

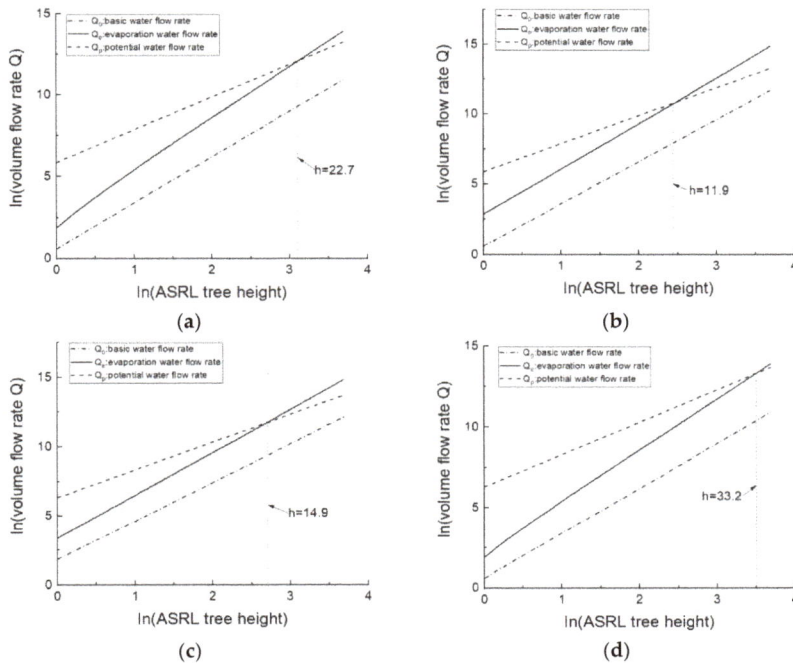

Figure 7. Analysis of optimization results. ASRL predictions of four case studies using verification point: (**a**) Model with parameters replacements, parametric optimizations, and topographic index; (**b**) Model with parametric optimizations and topographic index; (**c**) Model with parameters replacements and topographic; and (**d**) Model with parameters replacements and parametric optimizations. X axis represents the logarithm of tree height (unit: m), and Y axis represents the logarithm of water flow rate (unit: L/year).

5.2. Model Sensitivity

Sensitivity analysis presents the potential influence for predictions of the ASRL model by the climatic inputs, including precipitation, wind speed, vapor pressure, air temperature, and solar radiation. Changes in the water flow rates (Figure 8a–d) and maximum tree heights (Figure 8e–h) are investigated by perturbing each climatic variable while keeping the others constant. Intervals of variable alteration are 0.2 °C for temperature (ranging from −2 °C to 2 °C) and 2% for others (ranging from −20% to 20%). The monthly climatic variables of the verification point (121.554° E, 53.291° N) imported to the ASRL model are shown in Table 2.

As Figure 8 shows, the modeled water flow rates and potential maximum tree heights are sensitive to changes of climatic variables, and the direction and magnitude of model sensitivity are not the same across different variables. For instance, the potential water flow rate is influenced by precipitation, while the evaporation flow rate is sensitive to the others. A 20% increase in precipitation (Figure 8a) and vapor pressure (Figure 8e) produced a greater maximum tree height prediction (Δh_{max} = 3.9 m, Δh_{max} = 10.9 m). The modeled maximum tree heights are positively correlated with precipitation (Figure 8e) and vapor pressure (Figure 8f). In contrast, the predicted maximum tree height became smaller (Δh_{max} = −3.4 m, Δh_{max} = −5.7 m) when wind speed (Figure 8b) and air temperature (Figure 8d) were added, meaning a negative correlation between modeled tree height and the two variables. Comparing the slopes of the four curves in Figure 8e–h, the ASRL modeled maximum tree height is more sensitive to changes in vapor pressure (Figure 8g) and air temperature (Figure 8h) than precipitation (Figure 8e) and wind speed (Figure 8f). Changes in wind speed and vapor pressure show contrary patterns of magnitude sensitivity.

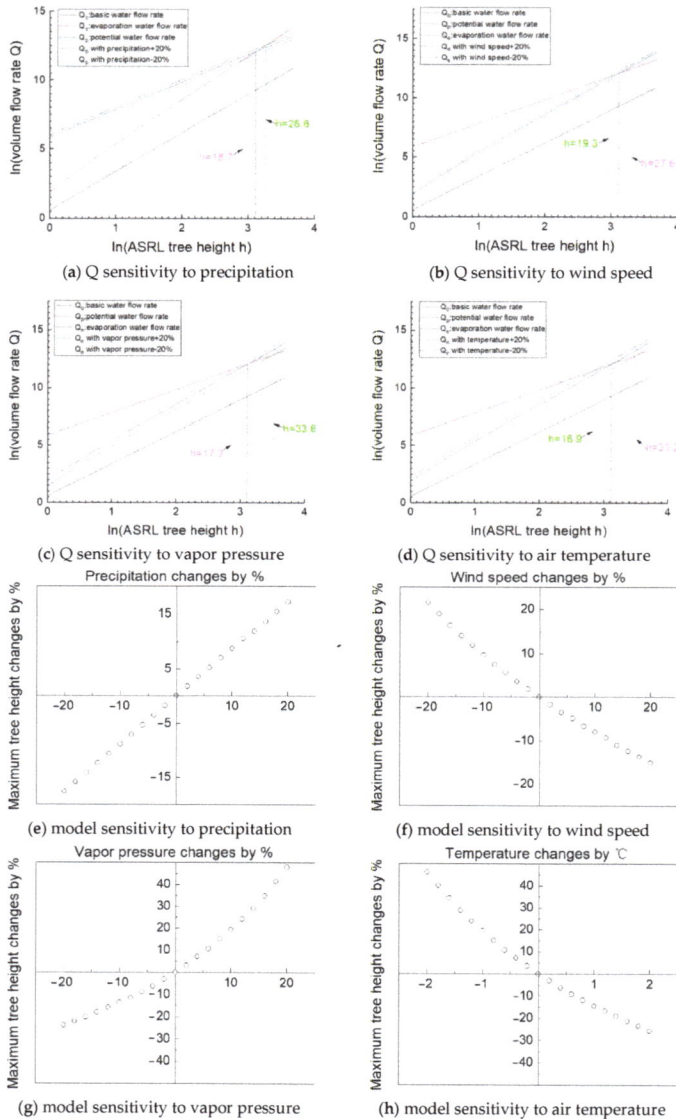

(a) Q sensitivity to precipitation

(b) Q sensitivity to wind speed

(c) Q sensitivity to vapor pressure

(d) Q sensitivity to air temperature

(e) model sensitivity to precipitation

(f) model sensitivity to wind speed

(g) model sensitivity to vapor pressure

(h) model sensitivity to air temperature

Figure 8. Sensitivity analysis of the ASRL model. The sensitivity to climatic variables including precipitation, wind speed, vapor pressure, and temperature. (**a–d**) Changes in the water flow rates are investigated by perturbing each climatic variable while keeping the others constant (precipitation, wind speed, and vapor pressure changed by ±20%, while temperature changed by ±2 °C). (**e–h**) Percent changes in maximum tree heights are investigated by perturbing each climatic variables while keeping others constant (precipitation, wind speed, and vapor pressure changed from −20% to 20% at a rate of 2%, while temperature changing from −2 °C to 2 °C at a rate of 0.2 °C).

Table 2. The monthly climatic inputs of the verification point.

Group	prcp	wnd	vp	tmp	srad
January	3	0.8	0.04	−28.7	4.13
February	4	1	0.06	−23.1	7.314
March	10	1.5	0.13	−13.6	12.055
April	23	2.1	0.3	−0.9	16.313
May	32	2.1	0.5	7.8	19.596
June	72	1.6	1.06	14.6	21.009
July	112	1.4	1.46	17.4	19.576
Auguest	102	1.3	1.28	14.8	16.104
September	49	1.5	0.68	7.1	11.953
October	19	1.5	0.29	−4.2	8.073
November	9	1	0.11	−19.1	4.649
December	5	0.7	0.05	−27.4	3.116
Unit	mm	m s^{-1}	kPa	°C	MJ m^{-2} day^{-1}

prcp, monthly total precipitation; wnd, mean wind speed; vp, mean vapor pressure; tmp, mean temperature; srad, mean solar radiation.

The ASRL model is least sensitive to solar radiation, similar to Choi's [27] result. For a 20% change in solar radiation, the predicted maximum tree height changes within 0.1 m. The reasons for this phenomenon are twofold: First, our research area is a water-limited environment, which means energy is not a major constraint on tree growth. Second, our study area belongs to a middle and high latitude region with low solar radiation. Due to the lack of experimental data, no more verification has been done.

6. Conclusions

In this paper, metabolic scaling theory and the Penman–Monteith equation are applied in the ASRL model to estimate maximum tree heights in the Greater Khingan Mountain, Inner Mongolia. Temperature, precipitation, wind speed, vapor pressure, and solar radiation are key input variables of the ASRL model. Model improvement and model sensitivity are also discussed to demonstrate the prognostic application of the ASRL model. Through our research, we found,

1. New values of the scaling coefficients ϕ and β_3 from field measurements make the model more consistent with the forest growth state of the study area.
2. Optimization of three parameters, β_1, γ, and s_{leaf}, improves the accuracy of the model prediction.
3. The introduction of a normalized topography index can effectively avoid overestimating short trees' heights in high slope areas and underestimating tall trees' heights in low slope areas.
4. Sensitivity analysis indicates the ASRL maximum tree height model is more sensitive to temperature and vapor pressure than any other climatic variables.

Caution is required in interpreting the results of the ASRL model because the current spatial scale fails to capture local tree height influenced by the small-scale climate variables, especially in mountains and valleys. Furthermore, species-specific parameters are not included in the model, which may also affect estimation of tree height and hence biomass. Thus, a progression of this work would be to account for application of high spatial resolution climate data and species information, and to assess the performance and utility of these techniques in other forests.

Author Contributions: Conceptualization, R.B.M., Y.S.; methodology, S.C., Y.S., Y.Z., X.N. and R.B.M.; investigation, Y.Z. and Y.S.; resources, X.N. and Y.Z.; writing-original draft preparation, Y.Z. and Y.S.; writing-review and editing, Y.Z. and Y.S.

Funding: This research was funded by National natural science foundation of China, grant number 41471312.

Conflicts of Interest: The authors declare no conflict of interest.

References

1. Clark, D.B.; Clark, D.A. Landscape-scale variation in forest structure and biomass in a tropical rain forest. *For. Ecol. Manag.* **2000**, *137*, 185–198. [CrossRef]
2. Drake, J.B.; Dubayah, R.O.; Clark, D.B.; Knox, R.G.; Blair, J.B.; Hofton, M.A.; Chazdon, R.L.; Weishampel, J.F.; Prince, S. Estimation of Tropical Forest Structural Characteristics Using Large-Footprint Lidar. *Remote Sens. Environ.* **2002**, *79*, 305–319. [CrossRef]
3. Muraoka, H.; Koizumi, H. Satellite Ecology (SATECO)—Linking ecology, remote sensing and micrometeorology, from plot to regional scale, for the study of ecosystem structure and function. *J. Plant Res.* **2009**, *122*, 3–20. [CrossRef] [PubMed]
4. Laumonier, Y.; Edin, A.; Kanninen, M. Landscape-scale variation in the structure and biomass of the hill dipterocarp forest of Sumatra: implications for carbon stock assessments. *For. Ecol. Manag.* **2010**, *259*, 505–513. [CrossRef]
5. West, G.B.; Enquist, B.J.; Brown, J.H. A general quantitative theory of forest structure and dynamics. *Proc. Natl. Acad. Sci. USA* **2009**, *106*, 7040–7045. [CrossRef]
6. Enquist, B.J.; Niklas, K.J. Invariant scaling relations across tree-dominated communities. *Nature* **2001**, *410*, 655–660. [CrossRef]
7. Enquist, B.J.; Brown, J.H.; West, G.B. Allometric scaling of plant energetics and population density. *Nature* **1998**, *395*, 163–165. [CrossRef]
8. Muller-Landau, H.; Condit, R.; Harms, K.; Marks, C.; Thomas, S. Comparing tropical forest tree size distributions with the predictions of metabolic ecology and equilibrium models. *Ecol. Lett.* **2006**, *9*, 589–602. [CrossRef]
9. Enquist, B.J.; West, G.B.; Brown, J.H. Extensions and evaluations of a general quantitative theory of forest structure and dynamics. *Proc. Natl. Acad. Sci. USA* **2009**, *106*, 7046–7051. [CrossRef] [PubMed]
10. Pang, Y.; Zhao, F.; Li, Z. Forest Height Inversion using Airborne Lidar Technology. *J. Remote Sens.* **2008**, *12*, 152–158.
11. Xiyun, M.; Qiuliang, Z. Inversion of Forest Height and Canopy Closure Using Airborne LiDAR Data. *J. Northeast For. Univ.* **2015**, *43*, 84–89.
12. Lefsky, M.A. A global forest canopy height map from the Moderate Resolution Imaging Spectroradiometer and the Geoscience Laser Altimeter System. *Geophys. Res. Lett.* **2010**, *37*, 78–82. [CrossRef]
13. Simard, M.; Pinto, N.; Fisher, J.B. Mapping forest canopy height globally with spaceborne lidar. *J. Geophys. Res. Biogeosci.* **2015**, *116*, 4021. [CrossRef]
14. Xuehui, L.; Ainong, L.; Guangbin, L. An Approach to Decompose ICESat/GLAS Data Waveform and Estimate Canopy Height Based on PSO-LSM Method. *Geogr. Geo-Inform. Sci.* **2017**, *33*, 22–29.
15. Liang, Z.; Xiaoqi, J.; Weiwei, Z. Retrieval of forest canopy height based on large-footprint LiDAR data. *Sci. Surv. Map.* **2018**, *43*, 148–153.
16. Hong, C.; Jinliang, H.; Juan, Q. Estimation of forest aboveground biomass using ICESat/GLAS data and Landsat/ETM+ imagery. *Sci. Surv. Map.* **2018**, *43*, 9–16.
17. Goetz, S.; Dubayah, R. Advances in remote sensing technology and implications for measuring and monitoring forest carbon stocks and change. *Carbon Manag.* **2011**, *2*, 231–244. [CrossRef]
18. Tang, H.; Dubayah, R.; Brolly, M. Large-scale retrieval of leaf area index and vertical foliage profile from the spaceborne waveform lidar (GLAS/ICESat). *Remote Sens. Environ.* **2014**, *154*, 18. [CrossRef]
19. Stojanova, D.; Panov, P.; Valentin, G.B. Estimating Vegetation Height and Canopy Cover from Remotely Sensed Data with Machine Learning. *Ecol. Inform.* **2010**, *5*, 256–266. [CrossRef]
20. Kempes, C.P.; West, G.B.; Crowell, K. Predicting Maximum Tree Heights and Other Traits from Allometric Scaling and Resource Limitations. *PLoS ONE* **2011**, *6*, E20551. [CrossRef]
21. Shi, Y.; Ni, X.; Choi, S. Allometric Scaling and Resource Limitations Model of Tree Heights: Part 1. Model Optimization and Testing over Continental USA. *Remote Sens.* **2014**, *6*, 284–306. [CrossRef]
22. Choi, S.; Ni, X.; Shi, Y. Allometric Scaling and Resource Limitations Model of Tree Heights: Part 2. Site Based Testing of the Model. *Remote Sens.* **2013**, *5*, 202–223. [CrossRef]
23. Ni, X.; Park, T.; Choi, S. Allometric Scaling and Resource Limitations Model of Tree Heights: Part 3. Model Optimization and Testing over Continental China. *Remote Sens.* **2014**, *6*, 284–306. [CrossRef]

24. Nemani, R.R. Climate-Driven Increases in Global Terrestrial Net Primary Production from 1982 to 1999. *Science* **2003**, *300*, 1560–1563. [CrossRef]

25. Zhuoting, W.; Dijkstra, P.; Koch, G.W. Responses of terrestrial ecosystems to temperature and precipitation change: A meta-analysis of experimental manipulation. *Glob. Chang. Biol.* **2011**, *17*, 927–942.

26. Peng, S.; Piao, S.; Ciais, P. Asymmetric effects of daytime and night-time warming on Northern Hemisphere vegetation. *Nature* **2013**, *501*, 88–92. [CrossRef]

27. West, G.B.; Brown, J.H.; Enquist, B.J. A general model for the origin of allometric scaling laws in biology. *Science* **1997**, *276*, 122–126. [CrossRef]

28. Monteith, J.L.; Unsworth, M.H. Principles of environmental physics: Plants, animals, and the atmosphere. *Acad. Press Oxf.* **2013**, *4*, 217–247.

29. Choi, S.; Kempes, C.P.; Park, T. Application of the metabolic scaling theory and water–energy balance equation to model large-scale patterns of maximum forest canopy height. *Glob. Ecol. Biogeogr.* **2016**, *25*, 1428–1442. [CrossRef]

30. Cao, C.; Ni, X.; Wang, X. Allometric scaling theory-based maximum forest tree height and biomass estimation in the Three Gorges reservoir region using multi-source remote-sensing data. *Int. J. Remote Sens.* **2016**, *37*, 13. [CrossRef]

31. Kempes, C.P.; Dutkiewicz, S.; Follows, M.J. Growth, metabolic partitioning, and the size of microorganisms. *Proc. Natl. Acad. Sci. USA* **2012**, *109*, 495–500. [CrossRef]

32. Friedl, M.A.; Sullamenashe, D.; Tan, B. MODIS Collection 5 global land cover: algorithm refinements and characterization of new datasets. *Remote Sens. Environ.* **2017**, *114*, 168–182. [CrossRef]

33. Meinzer, F.C. Functional Convergence in Plant Responses to the Environment. *Oecologia* **2003**, *134*, 1–11. [CrossRef]

34. Schmidtnielsen, K. *Scaling: Why Is Animal Size so Important?* Cambridge University: Cambridge, UK, 2002.

35. Brown, J.H.; West, G.B.; Enquist, B.J. A general model for the structure and allometry of plant vascular systems. *Nature* **1999**, *400*, 664–667.

36. Niklas, K.J.; Spatz, H.C. Growth and hydraulic (not mechanical) constraints govern the scaling of tree height and mass. *Proc. Natl. Acad. Sci. USA* **2004**, *101*, 15661–15663. [CrossRef]

37. Niklas, K.J. Maximum plant height and the biophysical factors that limit it. *Tree Physiol.* **2007**, *27*, 433–440. [CrossRef]

38. Pretzsch, H.; Dieler, J. Evidence of variant intra- and interspecific scaling of tree crown structure and relevance for allometric theory. *Oecologia (Berlin)* **2012**, *169*, 637–649. [CrossRef]

39. Lin, Y.; Berger, U.; Grimm, V. Plant Interactions Alter the Predictions of Metabolic Scaling Theory. *PLoS ONE* **2013**, *8*. [CrossRef]

40. Duncanson, L.I.; Dubayah, R.O.; Enquist, B.J. Assessing the general patterns of forest structure: Quantifying tree and forest allometric scaling relationships in the United States. *Glob. Ecol. Biogeogr.* **2015**, *24*, 1465–1475. [CrossRef]

41. Smith, D.D.; Sperry, J.S.; Enquist, B.J. Deviation from symmetrically self-similar branching in trees predicts altered hydraulics, mechanics, light interception and metabolic scaling. *New Phytol.* **2014**, *201*, 217–229. [CrossRef]

forests

MDPI

Article

Can Field Crews Telecommute? Varied Data Quality from Citizen Science Tree Inventories Conducted Using Street-Level Imagery

Adam Berland [1,*], **Lara A. Roman** [2] **and Jess Vogt** [3]

[1] Department of Geography, Ball State University, 2000 W University Ave, Muncie, IN 47306, USA
[2] US Forest Service, Northern Research Station, Philadelphia Field Station, 100 N. 20th St. Suite 205, Philadelphia, PA 19103, USA; lara.roman@usda.gov
[3] Environmental Science and Studies Department, DePaul University, 1 E. Jackson Blvd., Chicago, IL 60601, USA; jess.vogt@depaul.edu
*   Correspondence: amberland@bsu.edu

Received: 18 March 2019; Accepted: 18 April 2019; Published: 20 April 2019

check for updates

Abstract: Street tree inventories are a critical component of urban forest management. However, inventories conducted in the field by trained professionals are expensive and time-consuming. Inventories relying on citizen scientists or virtual surveys conducted remotely using street-level photographs may greatly reduce the costs of street tree inventories, but there are fundamental uncertainties regarding the level of data quality that can be expected from these emerging approaches to data collection. We asked 16 volunteers to inventory street trees in suburban Chicago using Google Street View[TM] imagery, and we assessed data quality by comparing their virtual survey data to field data from the same locations. We also compared virtual survey data quality according to self-rated expertise by measuring agreement within expert, intermediate, and novice analyst groups. Analyst agreement was very good for the number of trees on each street segment, and agreement was markedly lower for tree diameter class and tree identification at the genus and species levels, respectively. Interrater agreement varied by expertise, such that experts agreed with one another more often than novices for all four variables assessed. Compared to the field data, we observed substantial variability in analyst performance for diameter class estimation and tree identification, and some intermediate analysts performed as well as experts. Our findings suggest that virtual surveys may be useful for documenting the locations of street trees within a city more efficiently than field crews and with a high level of accuracy. However, tree diameter and species identification data were less reliable across all expertise groups, and especially novice analysts. Based on this analysis, virtual street tree inventories are best suited to collecting very basic information such as tree locations, or updating existing inventories to determine where trees have been planted or removed. We conclude with evidence-based recommendations for effective implementation of this type of approach.

Keywords: crowdsourced data; Google Street View; interrater agreement; municipal forestry; species identification; street trees; tree measurement; urban ecology; urban forestry

---

1. Introduction

Street trees are trees growing in the public right-of-way along streets in cities, towns, and suburbs. Though street trees are a relatively small proportion of the overall urban forest in many cities [1], they are a highly visible component that constitute a major focus of public engagement by municipalities and nonprofit organizations [2–4]. Street trees generate a range of benefits such as increased property values, shade, stormwater capture, and aesthetics [5–7]. Street trees can be conceived of as common pool resources, in that they provide benefits to the general public, but it is often unclear who is responsible for the care and maintenance these trees need to ensure future benefits provision [8]. The public right-of-way in which street trees are located is directly adjacent to both private and public properties, and may include a sidewalk. Although many municipalities have long directly managed street trees through municipal laws, policies, and personnel [2,9], these trees exist under a multi-stakeholder governance regime [10], with residents, municipal agencies, nonprofit organizations, and developers impacting decision-making.

For municipalities, there are clear costs to, first, not clearly defining management responsibilities for street trees and, second, not adequately monitoring and maintaining the trees themselves. If trees are not well maintained, they will not provide maximal benefits to the general public and in the worst instances may experience increased tree failure and mortality rates [11]. Furthermore, street trees require maintenance over time, they pose hazards when limbs or entire trees fall, and managers must balance considerations like tree condition and likelihood of limb or tree failure with the availability of time, money, and expertise to monitor trees and perform maintenance. Again, these decisions are further complicated by the multiple stakeholders invested in and gaining from street tree benefits and maintenance [8].

Informed, strategic management of street tree populations requires current inventory data [12]. Tree risk management – assessing mature trees for structural defects and taking appropriate actions such as pruning or removal – and associated concerns for liability drive municipal arborists to create and update street tree inventories [13,14]. Beyond maintenance of large street trees, managers also use inventories to make decisions about tree planting, with consideration for taxonomic and size class diversity, urban hardiness, maintenance requirements, and available resources [15–18]. Unfortunately, street tree inventory data are difficult to generate and to keep up-to-date. Inventories are typically conducted in the field by trained professionals, which is costly in terms of both time and money [12]. These costs preclude many municipalities from conducting an inventory in the first place, and prevent many others from updating their inventory data as often as they would like [19,20]. Inventory data are needed to effectively respond to crises like pest outbreaks, and furthermore, current inventory data open the door to more proactive management regimes [19,21–23]. In light of the high costs associated with field inventories conducted by trained professionals and limited municipal resources, there has been increased interest in using alternative means of data collection to generate street tree inventory data. This includes both citizen science field data and virtual surveys using emerging digital resources like Google Street View™ (Google Inc., Mountain View, CA, USA), as described below. The potential involvement of actors beyond municipal arborists in generating street tree inventories is important, given the multiple stakeholders invested in (and potentially contributing to the management of) street tree populations in the public right-of-way.

1.1. Citizen Science Tree Inventories

Citizen science refers to public participation in authentic research [24,25]. Citizen science permits more extensive data collection than could be accomplished by professional researchers alone, and it simultaneously increases public engagement with societally relevant scientific issues [25–27]. Crowdsourcing, in particular, engages a large number of people in data collection, and the term crowdsourcing is often used in reference to online data collection [28]. Internet-based mapping applications have facilitated the collection of geographically referenced citizen science data [24], and emerging technologies will continue to open new possibilities for the user experience and data management within citizen science projects [29]. Despite major increases in the size and scope of crowdsourcing projects (see examples at https://www.zooniverse.org), a fundamental question of data quality persists because volunteers lack the training and experience of experts when making scientific observations [30,31]. The perception that citizen science projects produce lower quality data limits the use of citizen science data in both practical management and peer-reviewed science publications. Assessments of citizen science data quality have had mixed results, with some studies demonstrating that volunteers can produce data of comparable quality to experts [26,31]. Notably, citizen science data quality varies depending in part on the task complexity, which reflects both the subject matter knowledge and intricacy of the data collection technique [31–33]. Effective training strategies can overcome some of the barriers to producing accurate citizen science data [26,33].

In light of the simultaneous need for street tree inventory data and the lack of municipal resources to conduct inventories, many communities have enlisted citizen scientists to generate street tree data [34,35]. However, only a few studies have formally assessed the quality of urban tree data produced by volunteers [26,27,32,36]. These efforts have shown that volunteers can produce reasonably high quality data (i.e., data of comparable quality to data collected by experts, or data that is deemed acceptable for the intended application). In general, citizen scientists performed best at recording more basic variables that require less expertise to document accurately, and their performance was poorer for more detailed variables and for variables requiring more subjective assessment. For example, volunteer data for genus identification was in agreement with expert identification over 90% of the time in three studies, while average agreement rates were lower (64%–85%) at the species level [26,27,32]. Similarly, citizen scientists recorded diameter at breast height (DBH) within 2.54 cm of the expert measurement roughly 90% of the time [26], but performance declined when a more precise standard of agreement with expert data was imposed. Finally, multiple studies have reported that citizen scientists struggle to produce data that is consistent with expert data for more subjective variables such as maintenance needs and tree condition ratings [26,27,32,36]. Overall, if the volunteers are asked to collect data within their capabilities, citizen science approaches to street tree inventories can be an effective way to generate data at a competitive cost while building community engagement and social capital in urban forestry programs [27].

1.2. Virtual Surveys Using Street-Level Imagery

Remotely sensed imagery has long been used to generate information about urban tree canopy cover [37–39], but these products are traditionally limited to canopy abundance information that does not distinguish individual trees or provide details like species composition or size class distribution. More recently, researchers have advanced the use of remote sensing to monitor individual trees [40], and to identify tree species remotely [41]. At present, these approaches require substantial computing expertise and expensive imagery that is not available in all areas. Publicly available imagery sources hold potential for generating street tree data widely and at a relatively lower cost compared to proprietary imagery. For example, Google Street View™ (GSV) (https://www.google.com/streetview/) provides street-level panoramas for most of the USA, and GSV coverage has expanded in recent years to include dozens of other countries. Similarly, Tencent Maps street view (http://map.qq.com) offers street-level panoramas in hundreds of Chinese cities.

Street-level panoramic photos capture trees lining streets, so several studies have explored the possibilities for using such imagery to generate information about street trees and streetscape greenery. Li, et al. [42] outlined an automated procedure to quantify green pixels in GSV images, and Li, et al. [43] applied this technique to study relationships between the abundance of street-level greenery and neighborhood socioeconomic characteristics. Tencent imagery was used to compare street greenery in 245 Chinese cities, and the results showed that cities in western China generally had greener streets than other regions [44]. GSV has been used to quantify streetscape shade provision in Singapore [45] and Boston [46]. Seiferling, *et al.* [47] used GSV imagery and innovative computer vision techniques to automatically detect the locations of trees and predict tree canopy cover. Computer vision was also used in Pasadena, California, to detect street trees, identify their species, and monitor changes over time [48]. While this approach successfully detected only about 70% of street trees, it showed great promise for identifying trees, monitoring changes, and potentially extending to additional metrics such as trunk diameter [48]. It seems likely that this type of approach, rooted in machine learning and computer vision, will eventually become a popular means of generating street tree data, because it is more efficient than human data collectors and data quality will improve as methods are refined. On the other hand, these sophisticated techniques for automated tree data generation are currently inaccessible to all but a very few communities simply because of the advanced computing skills required to implement the techniques.

In this study, we chose to explore simpler approaches to street tree data collection that are broadly accessible because they can be implemented by less skilled computer users. One such approach is a so-called virtual survey in which analysts record tree data by manually interpreting photos as if they were walking down the street [49]. This approach is similar to windshield surveys, which involve rapid data collection by a crew driving a vehicle along streets and have a long history in urban forestry practice and research [50,51]. Berland and Lange [49] compared GSV virtual surveys to field data from the same locations, and found that the analyst documented 93% of the trees inventoried in the field, and produced genus identification data that agreed with the field data for 90% of trees. While species identifications and diameter class estimates agreed with the field data less often (66% and 67%, respectively), the authors concluded that this simple approach to virtual surveys in GSV showed promise for generating street tree data efficiently and with a reasonable degree of reliability for basic variables [49]. However, the analyst in that study had a college degree in field botany and work experience in urban forestry, and communities interested in applying this approach may not have analysts on hand with such expertise. As such, we present here an investigation of the level of data quality that can be generated using GSV virtual surveys employing analysts ranging from trained experts to novice volunteers.

1.3. Aims and Research Questions

We are aware of several municipalities interested in using citizen science and/or virtual survey techniques to generate street tree inventory data [34,52–54]. However, there are fundamental uncertainties about the level of data quality that can be generated using a pairing of volunteer-generated data and virtual survey techniques. In this study, we compared virtual survey data to field data from the same locations as a means of assessing virtual survey data quality. We also assessed the level of agreement among analysts at three different self-rated expertise levels (novice, intermediate, and expert). We posed the following research questions:

1. To what degree do virtual survey analysts in the same self-rated expertise category agree with one another? Does this level of agreement vary among expertise categories? For example, do experts agree with one another more often than novices agree with one another?
2. What is the level of agreement between virtual survey data and field data? How does this vary according to analyst expertise?

Based on prior citizen science field studies for both urban and rural forestry [26,27,32,34,36,52,53,55,56], we anticipated that novices would produce high quality data for simple variables such as tree counts, but experts would produce higher quality data for variables such as species identification that require more background knowledge. Our findings provide insights into the applicability of using virtual surveys to generate street tree inventory data, and point to practical recommendations that can be used by urban forest managers considering this approach to data collection in their communities.

2. Materials and Methods

2.1. Study Area

The study area is the Village of Dolton, IL, USA (41.64° N, 87.61° W), which lies immediately south of Chicago. Dolton covers a land area of 12.1 km², and has approximately 111 km of local public roads. Dolton had a 2016 population of 23,091 [57]. The median household income was $44,075, and 27% of individuals lived below poverty level. 86% of residents had at least a high school degree, while 17% had a bachelor's degree or higher. The population was 91% black or African American, 7% white, and 4% identified as Hispanic or Latino (of any race). Most homes (77%) were built between 1950 and 1979, 64% of residences were owner-occupied, and the median home value was $94,700 [57].

Urban forest management in Dolton is constrained by limited municipal resources. There is no formal urban forestry department or certified arborist in Dolton dedicated to managing street trees. Street trees in Dolton are under the jurisdiction of the Public Works Department, which maintains all public infrastructure including streets, sidewalks, and water/sewer lines [58]. However, there is little active street tree planning and management from municipal personnel. We chose to work in Dolton because the village had expressed interest in acquiring street tree data to our contacts at the Morton Arboretum, an organization that supports tree science and stewardship in the Chicago region. Along streets in Dolton, most trees appear to have been planted at or around the time of housing construction (estimated age 40–60 years old) or possibly planted later by a homeowner.

2.2. Field Data Collection

The field crew for data collection was composed of two senior undergraduate students majoring in Environmental Studies at DePaul University. The students had tree identification skills from applied ecology field work courses that were strengthened during three full days of urban forest inventory training with the coauthor supervising field work. Field data were collected during summer 2017. The field crew completed a random sample of street segments along Dolton's local public roads. Forty-three street segments were used in the analysis described below to match virtual survey sampling effort. Field data were collected digitally using ESRI's Survey123 (ESRI, Redlands, CA, USA). For each tree encountered, the field crew recorded location information including the street segment identifier, the street address, and a sequential number to differentiate between multiple trees at a given street address. They documented tree information including DBH to the nearest 0.25 cm (0.1 inches), mortality status (alive, standing dead, or stump), genus, species, special notes (e.g., description of the tree's location when the street address was not evident), and a timestamp for the record. The field crew also took multiple photos of each tree including the entire tree, close-ups of the leaves and bark, and any defining characteristics to help with species identification. These photos were used by the authors to confirm species identification when the field crew was uncertain.

2.3. Google Street View Virtual Surveys

2.3.1. Recruitment and Training of Analysts

Virtual survey analysts were recruited via email. Potential analysts were identified using the Morton Arboretum's contacts in greater Chicago and the professional networks of the authors. Individuals who were interested in participating were sent a link to a Google Forms™ survey that was used to obtain informed consent to participate and to collect self-rated expertise information (Supplementary Material S1). Respondents were asked about their previous experience in citizen science projects, experience and awareness regarding urban forestry management and field techniques (including tree measurement and identification), and experience and comfort using Google Maps™ and GSV. They were also asked background questions about their education, field of employment, and age. We sought a minimum of three analysts per expertise group, where experts were defined as those with substantial experience in urban forestry in their professional work, intermediate analysts were those with some experience collecting tree data in the field and moderate confidence in measuring and identifying trees, and novice users had very little or no experience collecting tree data in the field and low confidence in their ability to measure and identify trees.

Citizen science projects typically involve a training component to familiarize volunteers with field techniques [26,32,36]. Our study was not conducted face-to-face, so the virtual survey analysts received a set of digital training materials. Each analyst received a PDF document as an email attachment (Supplementary Material S2). This PDF document contained a project overview; illustrated definitions of key terms such as tree, street tree, public right-of-way, and street segment; instructions for collecting data; links to instructional YouTube™ (YouTube LLC, San Bruno, CA, USA) videos demonstrating data collection procedures; links to illustrated reference guides for estimating tree diameter and identifying species; and links to the analyst's personalized data collection form. This training document also contained contact information for the primary investigator in case the analyst had questions or problems while collecting data.

2.3.2. Data Collection Procedure

Virtual survey analysts collected data along the same list of randomly drawn street segments in Dolton. Street segments were defined as the portion of a street between two intersections. To inventory a street segment, analysts clicked on a link to the segment in a PDF document that directed them to the beginning of that segment in GSV. The street segment list named the intersecting streets and the address range along that segment to clarify the precise extent of that segment. Analysts navigated down the left side of the street in GSV, noting details about any trees they encountered (see below), or recording that there were no trees present along the left side of the segment. Then they returned to the street segment list, clicked that same link again, and repeated the procedure along the right side of the street. This process of inventorying the left and right sides of the street separately reduced the likelihood of user error. It also helped the authors match trees across multiple users for analysis purposes because trees were inventoried in a clearly defined order.

When analysts encountered a tree, they recorded information about the tree in a personalized data collection form shared privately online between the analyst and the authors on Google Sheets^{TM}. For each tree, the analyst noted the street segment number and side of the street (left or right), the street address number given in GSV, the estimated diameter class of the tree, the mortality status (alive, standing dead, or stump), genus, species, and identification confidence. Dropdown lists were used for street segment number, diameter class, mortality status, genus, species, and identification confidence to reduce data inconsistencies such as misspellings. The Google Sheet also contained a custom timestamp that automatically populated a start time when the analyst began recording information about the tree, and an end time when the analyst completed all the required fields for that tree. Diameter classes followed common bins used in urban forestry in the USA (0–7.6, 7.6–15.2, 15.2–30.5, 30.5–45.7, 45.7–61.0, 61.0–76.2, and >76.2 cm DBH). We did not ask analysts to estimate DBH with greater precision, because this proved difficult in a previous GSV study [49], and greater level of precision is not necessarily useful for most management purposes [32]. Analysts received a DBH reference guide that showed GSV images of trees within each diameter class, and links to those trees so the analyst could see what a tree of that size looked like in the native GSV imagery (Supplementary Material S3). The genus field contained a dropdown list of 51 common genera in the Chicago area drawn from an existing tree inventory [59], and analysts could select 'other' and type in a genus as appropriate. To aid in identification, analysts received a tree identification guide in PDF format that contained pictures and links to more information for approximately 60 common urban tree species from greater Chicago (Supplementary Material S4). Identification confidence was classified as high confidence, somewhat confident, or not confident.

2.4. Data Analysis

2.4.1. Time Comparison between Field and Virtual Surveys

We calculated the time taken to inventory trees in the field and in the virtual surveys. For the field data, we used timestamps from the tree records in Survey123 to break the field work into sampling blocks, where any gap between two trees greater than 30 minutes was considered a separate sampling block. This was designed to count short breaks during the field work, unique or difficult to identify trees that might take longer than average time to identify, and travel time between street segments, but not to count longer discontinuities such as lunch breaks against the total sampling time. After determining the length of each sampling block and the number of trees inventoried, we divided minutes by trees inventoried to get sampling time per tree. We did not include travel time to and from the study area in this calculation, because we are less interested in total time dedicated to field inventory activities than in per-tree comparison with virtual survey data collection. Time calculations for virtual survey data were based on Google Sheets timestamps. Again, we broke the data into sampling blocks defined by a gap less than 30 minutes between consecutive trees, and divided to calculate sampling time per tree.

2.4.2. Agreement within and among Analyst Expertise Groups

To assess the quality of data generated using virtual surveys, we compared virtual survey data among virtual survey analysts, and then separately compared virtual survey data to field data. When comparing virtual survey data among analysts, we assumed that higher agreement among analysts signaled higher data quality, particularly in terms of consistency or reliability. Of course, two analysts in agreement on a species identification could both be wrong, but we expected that agreement would more likely signal that both analysts arrived at the same accurate classification. Note that this is analogous to the standard means of assessing data quality in citizen science projects in the field, where the quality of citizen science data is judged against expert data, where the expert data are assumed to be correct even though they almost certainly contain some error [32].

Agreement among virtual survey analysts focused on the following four variables: tree count (the number of trees on each side of each street segment), diameter class, genus, and species. Analysts were instructed not to record species for *Amelanchier* Medik., *Cornus* L., *Crataegus* L., *Malus* Mill., and *Prunus* L., because these genera contain many hybrids and cultivars that can be difficult to distinguish [32]; for these genera, analysts were considered in agreement for species if they both selected the same genus. Together, these genera only comprised 1.5% of the total trees encountered, so this decision is likely to have a very minor impact on our results.

We tested agreement among all analysts by variable to determine which variables were collected more or less consistently among analysts. Raw percent agreement is biased by chance agreement, particularly when some classification categories are overrepresented. For example, in this study an analyst could have achieved 73% accuracy for genus identification by simply selecting maple (*Acer* L.) for every tree. Thus, we used Krippendorff's alpha statistic to measure agreement while correcting for chance agreement. Krippendorff's alpha ranges from 0 (no agreement) to 1 (perfect agreement). While there are no absolute benchmarks for satisfactory alpha values, scores >0.8 are typically considered to indicate good agreement, and scores >0.67 indicate acceptable agreement [60]. In this study, Krippendorff's alpha was more appropriate than other interrater reliability statistics because it allows for more than two raters. Furthermore, the statistic accommodates nominal, ordinal, interval, and ratio data [61], whereas other metrics only accommodate nominal data.

We implemented Krippendorff's alpha in R using the kripp.boot function [62]. We bootstrapped 95% confidence intervals using 10,000 iterations. Tree count was processed as interval data, DBH size class was processed as ordinal data, and genus and species were processed as nominal data. The implication of processing variables as ordinal or interval data is that raters are penalized more severely when ratings are in stronger disagreement. For genus and species identification, analysts were allowed to select 'unknown' when they felt unable to identify a tree. When calculating Krippendorff's alpha, each instance of an 'unknown' record was converted to its own unique genus or species code. This penalized users for selecting the unknown option, and provides a more conservative estimate of interrater agreement than other strategies (e.g., omitting unknowns from the analysis). We used Krippendorff's alpha (a) to examine differences among the four key tree variables for all users to evaluate which variables exhibited higher overall agreement, and (b) to compare within-group agreement across the three expertise groups for each of the four key variables to evaluate whether agreement levels varied among expertise groups. We interpreted non-overlapping confidence intervals as a conservative indicator of significant differences.

2.4.3. Agreement with Field Data

Next we compared data from each analyst to the field data using Krippendorff's alpha. Field measurements of DBH were converted from cm to diameter classes for comparison. Relatively higher Krippendorff's alpha scores indicated higher agreement between the analyst and the field data. We interpreted higher alpha scores as an indication of higher data quality, which assumes that the field data are correct. Notably, we could not account for tree plantings or removals between GSV imagery capture and field data collection. While field data were collected in summer 2017, virtual surveys were based on GSV imagery captured predominantly in 2012 and during summer months (Table 1).

Table 1. Year and month of Google Street View imagery availability for the street segments included in the virtual survey.

Year	% of Total	Month	% of Total
2009	3	June	< 1
2011	33	July	38
2012	51	August	55
2014	1	September	2
2015	7	November	5
2016	1		
2017	5		

3. Results

3.1. Overview of Virtual Survey Analysts and Tree Data

We recruited 16 analysts to participate in the virtual surveys. Their ages ranged from 22–72 years (median = 44.5 years). Nine had prior citizen science experience, and 15 of 16 analysts reported using Google Maps or Google Earth[TM] one or more times per week, but responses were more varied for Google Street View use. The analysts varied in their self-reported knowledge in urban forestry and confidence in their ability to complete tree identification and measurement tasks. Based on these responses, analysts were divided into expert, intermediate, and novice groups containing 3, 9, and 4 members, respectively.

The virtual survey analysts inventoried between 5–43 street segments (mean = 21.8 segments), and ranged from 53–357 trees (mean = 186.9 trees). The average time per tree was lowest for experts and highest for novices (Table 2). Nearly half of the analysts (7 of 16) averaged less time per tree than the field crew, but note that the analysts operated individually while the field crew consisted of two members. The data from the virtual surveys and the field survey can be found in Supplementary Material S5.

Table 2. Comparison of time spent per tree during data collection. Group ranges are in parentheses for virtual survey analysts.

Analyst Group	Minutes Per Tree
All virtual survey analysts	3.01 (1.07–8.75)
Expert	1.45 (1.07–1.90)
Intermediate	3.41 (1.45–8.75)
Novice	4.23 (2.99–7.16)
Field crew [1]	3.14

[1] Field crew times do not include travel time to and from the study area or organization of equipment. The field crew included two people, whereas the virtual survey analysts worked alone.

3.2. Agreement among Analysts

Interrater agreement among all virtual survey analysts was highest for tree count, followed by DBH class and genus, and species agreement was lowest (Figure 1). Only tree count had a

Krippendorff's alpha score above the heuristic threshold of 0.8 indicating good agreement [60]. Within expertise groups, experts had the highest level of agreement for DBH and tree count, and experts and intermediate analysts had higher agreement than novices for genus and species identification (Figure 2). Novices exhibited relatively high interrater variability on tree count, and this appeared to be caused by one analyst with a low sample size and low alpha score (see rightmost tree count data point in Figure 3). Novices had particularly low Krippendorff's alpha scores for genus (0.29) and species identification (0.23).

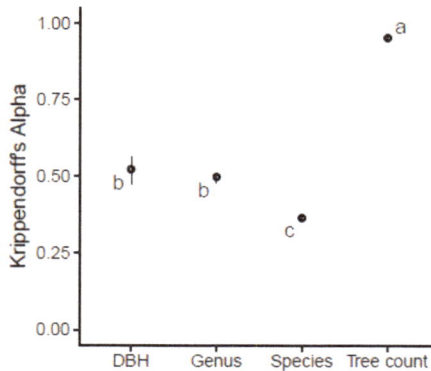

Figure 1. Krippendorff's alpha scores summarizing interrater agreement among all analysts for all variables, with 95% confidence intervals. Different letters indicate significantly different analyst agreement.

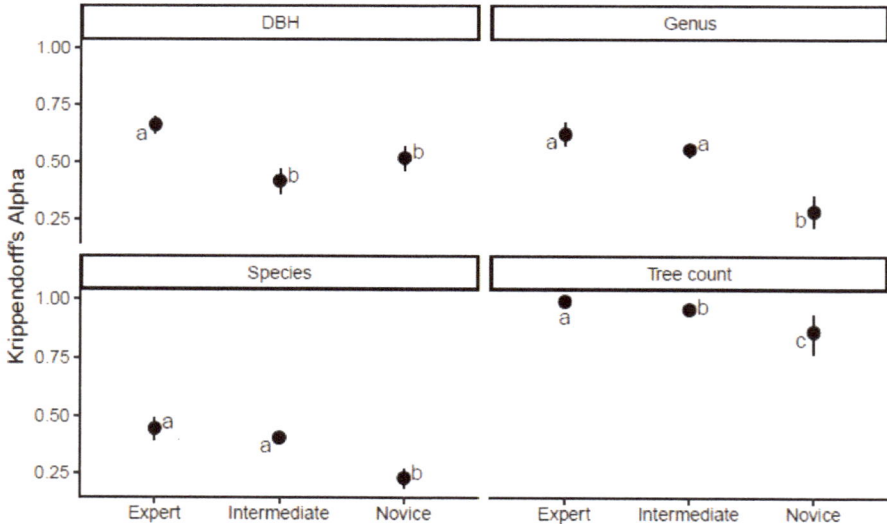

Figure 2. Krippendorff's alpha scores summarizing interrater agreement within analyst expertise groups by tree variable, with 95% confidence intervals. For each variable, different letters indicate significantly different analyst agreement across expertise groups.

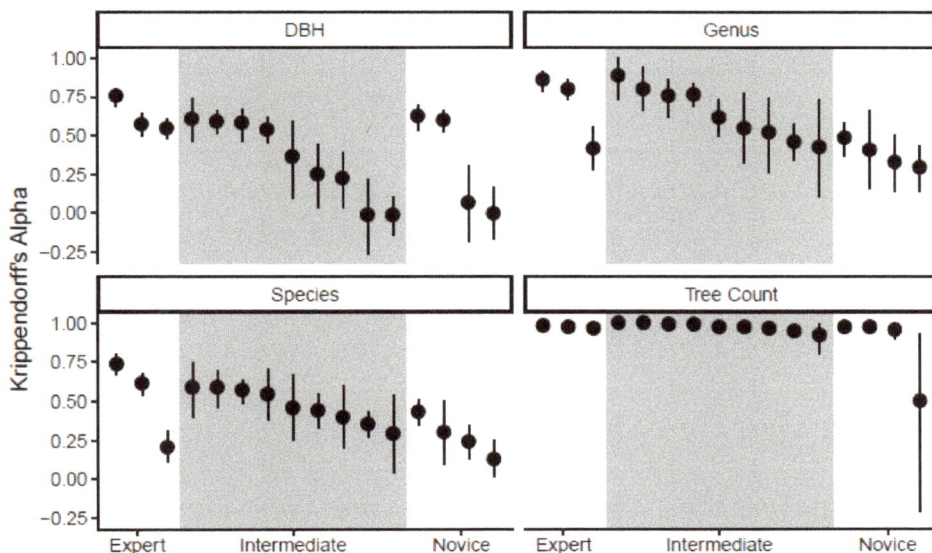

Figure 3. Krippendorff's alpha scores with 95% confidence intervals characterizing agreement between individual analysts and the field data.

3.3. Agreement between Field Data and Virtual Survey Data

Pairwise comparisons between virtual survey analysts and the field data show high agreement for 15 of 16 analysts for tree count (Figure 3). Two of the experts outperformed all of the novices for both genus and species identification; the third expert was penalized for selecting 'unknown' relatively frequently, resulting in a diminished alpha score for genus and species identification (Figure 3). We observed marked variability among intermediate and novice analysts for DBH, compared with relatively consistent performance among experts (Figure 3). DBH performance generally decreased with increasing tree sizes, and all analyst groups tended to underestimate the DBH of larger trees (Table 3). The top intermediate analysts performed on par with the experts across all variables (Figure 3).

Maples comprised 72.6% of the trees encountered in the field on the street segments analyzed here (Table 4), and genus agreement rates were high (88%–96%) for maples across all expertise groups. Identification performance was much poorer for other common genera including ash (*Fraxinus* L.), elm (*Ulmus* L.), and linden (*Tilia* L.) (Table 4). With respect to self-rated confidence in tree identification, analysts across all levels of expertise were more confident identifying to the genus level compared to the species level (Table 5). When the analysts were confident, overall agreement rates between virtual survey data and field data were high for both genus (94.5%) and species (89.6%) (Table 5). Experts were more likely to be in agreement with field data when they were confident, followed by intermediate and novice analysts. Agreement rates were substantially lower at both the genus and species levels when analysts rated their identifications as somewhat confident or not confident (Table 5).

Table 3. Diameter at breast height (DBH) classification outcomes by percent, summarized by diameter class and analyst expertise. *Agree* indicates agreement with the field measurement, *Under* indicates the percent of observations for which the virtual survey analyst underestimated DBH relative to the field measurement, and *Over* indicates the percent of observations for which the virtual survey analyst overestimated DBH relative to the field measurement.

	All			Expert			Intermediate			Novice		
DBH class	Under	Agree	Over	Under	Agree	Over	Under	Agree	Over	Under	Agree	Over
0–7.6 cm (n = 1)	–	0	100	–	0	100	–	0	100	–	0	100
7.6–15.2 cm (n = 4)	33	55	12	20	70	10	48	48	5	18	55	27
15.2–30.5 cm (n = 14)	41	46	14	40	47	13	48	39	13	28	56	16
30.5–45.7 cm (n = 64)	34	49	17	40	48	12	33	42	25	27	65	9
45.7–61.0 cm (n = 103)	56	35	9	54	37	9	54	34	13	66	33	2
61.0–76.2 cm (n = 84)	73	23	4	61	33	7	75	22	3	85	14	1
76.2 + cm (n = 61)	80	20	–	74	26	–	84	16	–	80	20	–
All trees (n = 331)	60	32	8	56	37	8	61	29	10	64	32	4

Table 4. Genus identification summarized by analyst expertise. *NA* indicates the genus was not encountered by an analyst from that group.

Genus	Field Relative Frequency (%)	Percent Agreement with Field Identification			
		All	Expert	Intermediate	Novice
Apple (*Malus* Mill.)	0.6	75	75	100	25
Ash (*Fraxinus* L.)	5.5	55	78	58	10
Cedar (*Thuja* L.)	0.3	60	100	50	0
Cherry (*Prunus* L.)	0.6	60	50	50	100
Dogwood (*Cornus* L.)	0.3	13	50	0	0
Elm (*Ulmus* L.)	9.2	45	68	40	20
Ginkgo (*Ginkgo* L.)	0.6	43	67	0	100
Hackberry (*Celtis* L.)	0.3	33	67	25	0
Honeylocust (*Gleditsia* L.)	2.8	68	77	90	21
Linden (*Tilia* L.)	2.2	53	57	60	32
Magnolia (*Magnolia* L.)	0.6	50	100	0	NA
Maple (*Acer* L.)	72.6	94	95	96	88
Mulberry (*Morus* L.)	0.3	0	0	0	NA
Oak (*Quercus* L.)	0.9	93	100	83	100
Redbud (*Cercis* L.)	0.3	100	100	100	NA
Spruce (*Picea* A. Dietr.)	0.3	88	100	89	75
Sweetgum (*Liquidambar* L.)	0.6	100	100	100	NA
Sycamore (*Platanus* L.)	0.9	56	71	69	0
Tuliptree (*Liriodendron* L.)	0.9	72	100	50	83
All trees	100.0	84	89	87	74

Table 5. Percent of total ratings and percent agreement with field data by confidence level and expertise level for genus and species identification.

Expertise	Confident		Somewhat Confident		Not Confident	
	% of Ratings	% Agreement	% of Ratings	% Agreement	% of Ratings	% Agreement
Genus						
All analysts	75.9	94.5	13.4	66.3	10.7	36.4
Expert	77.0	98.7	10.8	77.0	12.2	38.8
Intermediate	80.8	94.1	10.4	66.1	8.8	46.3
Novice	64.8	88.9	22.8	59.9	12.4	19.5
Species						
All analysts	50.2	89.6	39.6	50.0	10.2	23.3
Expert	47.5	96.0	40.9	59.6	11.6	26.1
Intermediate	57.7	89.8	33.8	46.5	8.4	28.4
Novice	38.9	78.7	49.4	44.7	11.7	12.5

4. Discussion

In this study, we used publicly available GSV imagery to generate street tree inventory data. We explored how data quality varied from analysts in three self-rated expertise groups. This work advances our understanding of data quality associated with two approaches to data collection – citizen science projects and virtual surveys using street-level panoramas – that have potential to become increasingly prominent for street tree inventories. Below we discuss our findings, describe limitations of our study, and offer recommendations for future application of this technique.

4.1. Data Quality from GSV Virtual Surveys

The 16 analysts in this study were diverse in age, educational background, and self-rated expertise related to urban forestry and tree inventory, yet the analysts were able to complete the project tasks using digital training materials and online data entry. Leveraging familiar, user-friendly digital platforms like GSV and Google Sheets reduced the time needed to train volunteers on new software or field equipment. This type of online data crowdsourcing may be useful in similar situations, because it would allow managers to solicit a substantial amount of data remotely without coordinating the schedules of volunteers for fieldwork activities. The rate of data collection for an individual analyst was competitive with that of field crews observed here (Table 2) and in other studies of citizen science tree inventories [27,32,36]. Slower data collection for novices was likely attributable to more time spent on tree identification. Analysts were able to participate on their own schedules, regardless of the weather, from any computer with an internet connection, and seasonal timing of the surveys was irrelevant because GSV imagery is primarily captured during the growing season (Table 1). On the other hand, by eliminating site visits and removing the teamwork aspect of field inventories, we may have missed out on some of the benefits of citizen science fieldwork including community building and enhanced public support for urban forestry initiatives [26,27].

Agreement among virtual survey analysts was high for the simplest variable, tree count (Figure 1), which is consistent with previous citizen science studies indicating that volunteers perform on par with experts for the most basic types of observations [26,27,32]. Even though we simplified data collection into DBH classes rather than requiring DBH estimation with greater precision, interrater agreement for DBH was relatively low (Figure 1), and most analysts were in low to marginal agreement with field data for DBH (Figure 3, Table 3). Likewise, we observed fairly poor agreement for tree identification at the genus and species levels (Figures 1–3) after Krippendorff's alpha accounted for the overabundance of maples in the study area (Table 4). The analyst's level of confidence in their identification was a good indicator of data quality, particularly for expert and intermediate analysts (Table 5). Overall, the expert group had the highest interrater agreement across the four variables we tracked, but even expert performance only exceeded a Krippendorff's alpha value of 0.8 for tree count (Figure 2). That said, experts also showed promise for identifying trees to the genus level; two expert analysts were in high agreement (alpha >0.8) for genus identification (Figure 3), while the third expert was frequently penalized for selecting 'unknown' for genus and species. Some intermediate analysts produced virtual survey data on par with experts across all four variables (Figure 3), indicating that this approach need not be used by experts alone. As suggested by Roman, et al. [32], volunteers could be evaluated for species identification accuracy at the beginning of a project, as high-performing individuals may not consistently identify themselves as such.

4.2. Limitations

Several limitations to our study warrant consideration. When comparing virtual survey data to field data, we used the term agreement rather than accurate to acknowledge that the field data likely contained some errors, albeit presumably minimal. We assumed that agreement with field data represented higher data quality, reflecting the idea that field data are the traditional standard for data collection. Updates to GSV imagery were beyond our control. Eighty-seven percent of the GSV imagery in this study was collected five or more years before the field data to which it was compared (Table 1). GSV imagery for Dolton was updated in summer 2018, but this update occurred after our analysts had completed the virtual surveys. Along with outdated imagery, we also encountered some blurry images and, less frequently, portions of streets such as dead ends that were not covered by GSV imagery. Where imagery was outdated, it is possible that some disagreement between the field crew and virtual survey analysts was attributable to analyst error or tree change (planting or removal).

The virtual survey analysts worked at their own pace as their schedules permitted. To avoid bias among analysts, we did not provide midstream feedback to help analysts confirm or recalibrate their estimates. Berland and Lange [49] found that this type of feedback can improve data quality as an analyst becomes more experienced. Moreover, Berland and Lange [49] had their analyst conduct a trial run by surveying a small selection of trees in GSV and then visiting them in the field to help the analyst calibrate his observations; this was not logistically possible in the present study.

4.3. Recommendations

Based on our findings, we present the following recommendations for those who may be interested in combining citizen science and virtual surveys to generate street tree inventory data. Our recommendations echo insights from citizen science and crowdsourcing studies for other environments and taxa.

- Street tree inventories are used for a wide variety of purposes requiring varying levels of data accuracy and precision [32]. In deciding whether a citizen science virtual survey is appropriate for your purposes, *consider how the data will be used* and then evaluate whether virtual survey data can reasonably be expected to meet those needs. Our study indicates that virtual surveys may provide useful information for management purposes, but perhaps not for research applications requiring more accurate and detailed data, similar to the findings from field-based street tree volunteer inventories in Roman, et al. [32]. Kosmala, *et al.* [63] recommend that citizen science data should be of sufficient quality to address management applications or research questions, implying that some volunteer data may be "good enough" for specific uses [52].
- *Task complexity should be tailored to the expertise of the analyst.* Novices may only be able to provide high quality data for basic variables like tree counts, street address, and broad size classes (e.g., small, medium, and large), while experts can produce more reliable data for more detailed tree attributes like genus [49]. When designing a study, familiarity with the potential pool of participants can help guide the level of task complexity to assign. An iterative process of study design and refinement should consider the data quality needed for the project objectives as well as the observed capabilities of volunteers [63].

- For species identification, volunteers' *self-reported confidence level* can be reported with each tree, or their overall *identification skills can be evaluated with pre-tests*. We observed substantially higher data quality when analysts were confident. Similarly, Roman, et al. [52] reported that crews had less confidence in the variables which showed the lowest consistency with experts. For virtual street tree surveys, it may be reasonable to accept the analyst's tree identification when they are confident (particularly for experts and intermediate analysts). However, data quality results regarding the importance of prior knowledge and self-reported confidence levels have been mixed in other citizen science contexts [30].

- *Virtual surveys should not replace field inspections by qualified professionals* for the purposes of identifying pest/pathogen problems (but see [64]). Similarly, virtual surveys are not appropriate for detailed assessments of tree risk because the analyst cannot examine the tree up close or from all angles, and because municipalities may want only certified arborists evaluating risk and pruning needs. However, for a very basic inventory that identifies standing dead trees in need of removal (as opposed to living trees that may be deemed structurally unstable or declining), virtual surveys may be appropriate [32,52,54].

- Virtual surveys *may not be suitable where in-person public engagement is a primary goal*. While virtual survey analysts would engage with the city's urban forestry program in the process of contributing data, virtual survey participants in general miss out on chances to engage with members of the public in the field [65], which occurs frequently as field crews inventory trees. For example, volunteers with a citizen science street tree inventory in New York City, NY were motivated by a desire to explore neighborhoods and meet new people [4]. However, virtual surveys may appeal to individuals who are physically disabled [66], and approaches such as gamification can deepen engagement and increase participation [67]. Future research about crowdsourcing in urban forestry could explore motivations for participation and the impacts of varied engagement approaches.

- *Consider how recent the imagery must be* to meet management needs. Street-level imagery products such as GSV offer nearly complete coverage of cities in the USA and many other countries, but imagery updates are not guaranteed on a timeline that is compatible with your needs. If the management goal is to produce a baseline inventory of street trees, then outdated imagery may be acceptable. But if the goal is to update an existing inventory, newer imagery would likely be necessary to characterize changes since the previous inventory. For example, a virtual survey could potentially be used to monitor trees from a planting program to assess whether the trees are alive, standing dead, or removed, as long as the GSV images are recent. Urban forestry nonprofits sometimes use volunteers to do such monitoring in the field [13,52], but the potential for online data collection could generate more data about planting program performance across towns that do not have the institutional capacity for field work.

- Implement a strategy to *provide midstream feedback to analysts*, as this can improve tree identification and estimation of DBH [49]. We provided analysts with illustrated examples of common tree species and GSV photos of trees in each DBH class, but analysts were not given feedback during the study to learn from their mistakes. Comparing their virtual survey data to field data for a subset of trees should help analysts refine the self-rated confidence and ultimately improve data quality. This strategy improved data quality in a previous study of virtual street tree surveys [49] and it aligns with feedback mechanisms designed to promote data quality and volunteer retention in other citizen science projects [63].

- *Rare taxa slow analysts down*, so we recommend using virtual surveys to efficiently inventory common species and note the locations of unidentifiable trees, which could be identified later in the field. Alternatively, the analyst could save screen captures of trees that are difficult to identify, and then an expert could attempt to identify the trees based on the photos. Such expert validation is a well-established strategy employed in citizen science projects from other fields [63,68]. Rare species have been shown to have particularly low species identification quality for other projects and taxa [69,70].

- *Use this technique to update existing street tree inventories.* In this study, analysts across all expertise groups excelled at documenting the locations of trees, but they struggled to identify trees to the species level and estimate DBH (Figures 2 and 3). Existing inventories will already contain information on tree species and sizes, so the analysts can focus on documenting tree plantings and removals. We expect this simple assessment of tree presence/absence would offer significant time savings over both field surveys and the more detailed virtual surveys used in this study.

5. Conclusions

Virtual surveys using street-level imagery offer an alternative or complementary approach to field data collection for street tree inventories. Based on data generated by 16 volunteer analysts using Google Street View™ imagery, we assessed data quality by comparing agreement among analysts in three expertise groups, and by comparing analyst data to field data from the same locations. We found that self-identified experts generally produced higher-quality data than intermediate and novice analysts, but some individual intermediate analysts produced data of comparable quality to experts. Data quality was high across all expertise groups for the simplest variable (tree counts), and data quality decreased for more detailed variables like species identification. However, when analysts were confident in their genus and species identifications, their data typically agreed with the field data. In practical terms, these findings point to the possibility of using virtual surveys to efficiently collect high-quality tree location data using novice volunteers, and the possibility of collecting more detailed tree data using more skilled analysts. Additional research and practical application in the areas of citizen science and virtual tree surveys will continue to improve our understanding of data quality associated with these approaches. With increasing interest in volunteer-generated data [53] as well as street-level imagery in urban forestry research and management [42–44,48,49], it is important to determine appropriate use cases and best practices for citizen science virtual surveys.

Supplementary Materials: The following are available online at http://www.mdpi.com/1999-4907/10/4/349/s1, Document S1: Questionnaire to collect self-rated expertise information and participant background data, Document S2: Training materials for virtual street tree surveys in Google Street View, Document S3: Tree diameter reference guide, Document S4: Tree identification reference guide, Document S5: Study data.

Author Contributions: Conceptualization, A.B., L.A.R., and J.V; methodology, A.B., L.A.R., and J.V.; formal analysis, A.B.; investigation, A.B., L.A.R., and J.V.; data curation, A.B. and J.V.; writing—original draft preparation, A.B.; writing—review and editing, L.A.R., J.V., and A.B.; visualization, A.B.; funding acquisition, A.B., L.A.R., and J.V.

Funding: This research and the APC were funded by the Tree Research and Education Endowment Fund (TREE Fund), grant number #16-JD-01.

Acknowledgments: This research was supported by the TREE Fund's John Z. Duling Grant (#16-JD-01). We are grateful to our volunteers for contributing virtual survey data, and to Kaitlyn Pike and Alli Preble for generating the field data. The Morton Arboretum guided us to select Dolton as a study area, and then helped us identify virtual survey participants.

Conflicts of Interest: The authors declare no conflict of interest. The funders had no role in the design of the study; in the collection, analyses, or interpretation of data; in the writing of the manuscript, or in the decision to publish the results.

References

1. Nowak, D.J.; Noble, M.H.; Sisinni, S.M.; Dwyer, J.F. People and trees: Assessing the US urban forest resource. *J. For.* **2001**, *99*, 37–42. [CrossRef]
2. Galenieks, A. Importance of urban street tree policies: A comparison of neighbouring Southern California cities. *Urban For. Urban Green.* **2017**, *22*, 105–110. [CrossRef]
3. Roy, S. Anomalies in Australian municipal tree managers' street-tree planting and species selection principles. *Urban For. Urban Green.* **2017**, *24*, 125–133. [CrossRef]

4.   Johnson, M.L.; Campbell, L.K.; Svendsen, E.S.; Silva, P. Why count trees? Volunteer motivations and experiences with tree monitoring in New York City. *Arboric. Urban For.* **2018**, *44*, 59–72.

5.   McPherson, G.; Simpson, J.R.; Peper, P.J.; Maco, S.E.; Xiao, Q. Municipal forest benefits and costs in five US cities. *J. For.* **2005**, *103*, 411–416.

6.   Mullaney, J.; Lucke, T.; Trueman, S.J. A review of benefits and challenges in growing street trees in paved urban environments. *Landsc. Urban Plan.* **2015**, *134*, 157–166. [CrossRef]

7.   Soares, A.L.; Rego, F.C.; McPherson, E.G.; Simpson, J.R.; Peper, P.J.; Xiao, Q. Benefits and costs of street trees in Lisbon, Portugal. *Urban For. Urban Green.* **2011**, *10*, 69–78. [CrossRef]

8.   Fischer, B.C.; Steed, B.C. Street trees—A misunderstood common-pool resource. In Proceedings of the 84th International Society of Arboriculture Annual Conference, St. Louis, MO, USA, 8 October 2008; p. 19.

9.   Ricard, R.M. Shade trees and tree wardens: Revising the history of urban forestry. *J. For.* **2005**, *103*, 230–233. [CrossRef]

10.   Mincey, S.K.; Hutten, M.; Fischer, B.C.; Evans, T.P.; Stewart, S.I.; Vogt, J.M. Structuring institutional analysis for urban ecosystems: A key to sustainable urban forest management. *Urban Ecosyst.* **2013**, *16*, 553–571. [CrossRef]

11.   Vogt, J.; Hauer, R.J.; Fischer, B.C. The costs of maintaining and not maintaining the urban forest: A review of the urban forestry and arboriculture literature. *Arboric. Urban For.* **2015**, *41*, 293–323.

12.   Östberg, J.; Delshammar, T.; Wiström, B.; Nielsen, A.B. Grading of parameters for urban tree inventories by city officials, arborists, and academics using the delphi method. *Environ. Manag.* **2013**, *51*, 694–708. [CrossRef] [PubMed]

13.   Roman, L.A.; McPherson, E.G.; Scharenbroch, B.C.; Bartens, J. Identifying common practices and challenges for local urban tree monitoring programs across the United States. *Arboric. Urban For.* **2013**, *39*, 292–299.

14.   Koeser, A.K.; Hauer, R.J.; Miesbauer, J.W.; Peterson, W. Municipal tree risk assessment in the United States: Findings from a comprehensive survey of urban forest management. *Arboric. J.* **2016**, *38*, 218–229. [CrossRef]

15.   Bond, J. *Best Management Practices—Tree inventories*; International Society of Arboriculture: Champaign, IL, USA, 2013; p. 35.

16.   Harris, R.; Clark, J.; Matheny, N. *Arboriculture: Integrated Management of Landscape Trees, Shrubs, and Vines*, 4th ed.; Prentice Hall: Upper Saddle River, NJ, USA, 2004; p. 592.

17.   McPherson, E.G.; Kotow, L. A municipal forest report card: Results for California, USA. *Urban For. Urban Green.* **2013**, *12*, 134–143. [CrossRef]

18.   Sjöman, H.; Östberg, J.; Bühler, O. Diversity and distribution of the urban tree population in ten major Nordic cities. *Urban For. Urban Green.* **2012**, *11*, 31–39. [CrossRef]

19.   Cowett, F.D.; Bassuk, N.L. Statewide assessment of street trees in New York State, USA. *Urban For. Urban Green.* **2014**, *13*, 213–220. [CrossRef]

20.   Maco, S.E.; McPherson, E.G. A practical approach to assessing structure, function, and value of street tree populations in small communities. *J. Arboric.* **2003**, *29*, 84–97.

21.   Ordóñez, C.; Duinker, P.N. An analysis of urban forest management plans in Canada: Implications for urban forest management. *Landsc. Urban Plan.* **2013**, *116*, 36–47. [CrossRef]

22.   Stobbart, M.; Johnston, M. A Survey of Urban Tree Management in New Zealand. *Arboric. Urban For.* **2012**, *38*, 247–254.

23.   Kenney, W.A.; van Wassenaer, P.J.E.; Satel, A.L. Criteria and indicators for strategic urban forest planning and management. *Arboric. Urban For.* **2011**, *37*, 108–117.

24.   Dickinson, J.L.; Shirk, J.; Bonter, D.; Bonney, R.; Crain, R.L.; Martin, J.; Phillips, T.; Purcell, K. The current state of citizen science as a tool for ecological research and public engagement. *Front. Ecol. Environ.* **2012**, *10*, 291–297. [CrossRef]

25.   Conrad, C.C.; Hilchey, K.G. A review of citizen science and community-based environmental monitoring: Issues and opportunities. *Environ. Monit. Assess.* **2011**, *176*, 273–291. [CrossRef] [PubMed]

26.   Bancks, N.; North, E.A.; Johnson, G.R. An analysis of agreement between volunteer- and researcher-collected urban tree inventory data. *Arboric. Urban For.* **2018**, *44*, 73–86.

27.   Bloniarz, D.V.; Ryan, H. The use of volunteer initiatives in conducting urban forest resource inventories. *J. Arboric.* **1996**, *22*, 75–82.

28. Estellés-Arolas, E.; Navarro-Giner, R.; González-Ladrón-de-Guevara, F. Crowdsourcing fundamentals: Definition and typology. In *Advances in Crowdsourcing*; Garrigos-Simon, F., Gil-Pechuán, I., Estelles-Miguel, S., Eds.; Springer: Cham, Switzerland, 2015; pp. 33–48.

29. Newman, G.; Wiggins, A.; Crall, A.; Graham, E.; Newman, S.; Crowston, K. The future of citizen science: Emerging technologies and shifting paradigms. *Front. Ecol. Environ.* **2012**, *10*, 298–304. [CrossRef]

30. Lewandowski, E.; Specht, H. Influence of volunteer and project characteristics on data quality of biological surveys. *Conserv. Biol.* **2015**, *29*, 713–723. [CrossRef]

31. Gillett, D.J.; Pondella, D.J.; Freiwald, J.; Schiff, K.C.; Caselle, J.E.; Shuman, C.; Weisberg, S.B. Comparing volunteer and professionally collected monitoring data from the rocky subtidal reefs of Southern California, USA. *Environ. Monit. Assess.* **2012**, *184*, 3239–3257. [CrossRef] [PubMed]

32. Roman, L.A.; Scharenbroch, B.C.; Östberg, J.P.A.; Mueller, L.S.; Henning, J.G.; Koeser, A.K.; Sanders, J.R.; Betz, D.R.; Jordan, R.C. Data quality in citizen science urban tree inventories. *Urban For. Urban Green.* **2017**, *22*, 124–135. [CrossRef]

33. Brandon, A.; Spyreas, G.; Molano-Flores, B.; Carroll, C.; Ellis, J. Can volunteers provide reliable data for forest vegetation surveys? *Nat. Areas J.* **2003**, *23*, 254–262.

34. Crown, C.A.; Greer, B.Z.; Gift, D.M.; Watt, F.S. Every Tree Counts: Reflections on NYC's Third Volunteer Street Tree Inventory. *Arboric. Urban For.* **2018**, *44*, 49–58.

35. Hauer, R.; Peterson, W. *Municipal Tree Care and Management in the United States: A 2014 Urban & Community Forestry Census of Tree Activities*; College of Natural Resources, University of Wisconsin-Stevens Point: Stevens Point, WI, USA, 2016; p. 71.

36. Cozad, S.; McPherson, E.G.; Harding, J.A. *STRATUM Case Study Evaluation in Minneapolis, Minnesota*; Center for Urban Forest Research, USDA Forest Service, Pacific Southwest Research Station: Davis, CA, USA, 2005.

37. Nowak, D.J.; Rowntree, R.A.; McPherson, E.G.; Sisinni, S.M.; Kerkmann, E.R.; Stevens, J.C. Measuring and analyzing urban tree cover. *Landsc. Urban Plan.* **1996**, *36*, 49–57. [CrossRef]

38. Rowntree, R.A. Forest canopy cover and land use in four Eastern United States cities. *Urban Ecol.* **1984**, *8*, 55–67. [CrossRef]

39. Zhang, Y. Texture-integrated classification of urban treed areas in high-resolution color-infrared imagery. *Photogramm. Eng. Remote Sens.* **2001**, *67*, 1359–1365.

40. Ossola, A.; Hopton, M.E. Measuring urban tree loss dynamics across residential landscapes. *Sci. Total Environ.* **2018**, *612*, 940–949. [CrossRef] [PubMed]

41. Alonzo, M.; Bookhagen, B.; Roberts, D.A. Urban tree species mapping using hyperspectral and lidar data fusion. *Remote Sens. Environ.* **2014**, *148*, 70–83. [CrossRef]

42. Li, X.; Zhang, C.; Li, W.; Ricard, R.; Meng, Q.; Zhang, W. Assessing street-level urban greenery using Google Street View and a modified green view index. *Urban For. Urban Green.* **2015**, *14*, 675–685. [CrossRef]

43. Li, X.; Zhang, C.; Li, W.; Kuzovkina, Y.A.; Weiner, D. Who lives in greener neighborhoods? The distribution of street greenery and its association with residents' socioeconomic conditions in Hartford, Connecticut, USA. *Urban For. Urban Green.* **2015**, *14*, 751–759. [CrossRef]

44. Long, Y.; Liu, L. How green are the streets? An analysis for central areas of Chinese cities using Tencent Street View. *PLoS ONE* **2017**, *12*, e0171110. [CrossRef] [PubMed]

45. Richards, D.R.; Edwards, P.J. Quantifying street tree regulating ecosystem services using Google Street View. *Ecol. Indic.* **2017**, *77*, 31–40. [CrossRef]

46. Li, X.; Ratti, C.; Seiferling, I. Quantifying the shade provision of street trees in urban landscape: A case study in Boston, USA, using Google Street View. *Landsc. Urban Plan.* **2018**, *169*, 81–91. [CrossRef]

47. Seiferling, I.; Naik, N.; Ratti, C.; Proulx, R. Green streets—Quantifying and mapping urban trees with street-level imagery and computer vision. *Landsc. Urban Plan.* **2017**, *165*, 93–101. [CrossRef]

48. Branson, S.; Wegner, J.D.; Hall, D.; Lang, N.; Schindler, K.; Perona, P. From Google Maps to a fine-grained catalog of street trees. *ISPRS J. Photogramm. Remote Sens.* **2018**, *135*, 13–30. [CrossRef]

49. Berland, A.; Lange, D.A. Google Street View shows promise for virtual street tree surveys. *Urban For. Urban Green.* **2017**, *21*, 11–15. [CrossRef]

50. Bassuk, N.L. Street tree diversity—Making better choices for the urban landscape. In Proceedings of the 7th Conference of the Metropolitan Tree Improvement Alliance, Lisle, IL, USA, 11–12 June 1990; pp. 71–78.

51. Rooney, C.J.; Ryan, H.D.P.; Bloniarz, D.V.; Kane, B.C.P. The reliability of a windshield survey to locate hazards in roadside trees. *J. Arboric.* **2005**, *31*, 89–94.

52. Roman, L.A.; Smith, B.C.; Dentice, D.; Maslin, M.; Abrams, G. Monitoring young tree survival with citizen scientists: The evolving Tree Checkers Program in Philadelphia, PA. *Arboric. Urban For.* **2018**, *44*, 255–265.

53. Roman, L.A.; Campbell, L.K.; Jordan, R.C. Civic science in urban forestry: An introduction. *Arboric. Urban For.* **2018**, *44*, 41–48.

54. Maldonado, S. Philly mapped street trees for smarter maintenance. *PlanPhilly*, 29 July 2016.

55. Butt, N.; Slade, E.; Thompson, J.; Malhi, Y.; Riutta, T. Quantifying the sampling error in tree census measurements by volunteers and its effect on carbon stock estimates. *Ecol. Appl.* **2013**, *23*, 936–943. [CrossRef] [PubMed]

56. Harrison, B.; Martin, T.E.; Mustari, A.H. The accuracy of volunteer surveyors for obtaining tree measurements in tropical forests. *Ambio* **2019**, 1–9. [CrossRef]

57. US Census Bureau. American FactFinder. Available online: http://factfinder.census.gov/ (accessed on 13 March 2019).

58. Village of Dolton. Public Works. Available online: https://vodolton.org/departments/public-works/ (accessed on 13 March 2019).

59. Nowak, D.J.; Hoehn, R.; Bodine, A.R.; Crane, D.E.; Dwyer, J.F.; Bonnewell, V.; Watson, G. *Urban Trees and Forests of the Chicago Region*; Resource Bulletin NRS-84; USDA Forest Service Northern Research Station: Newtown Square, PA, USA, 2013.

60. Krippendorff, K. Reliability in content analysis. *Hum. Commun. Res.* **2004**, *30*, 411–433. [CrossRef]

61. Hayes, A.F.; Krippendorff, K. Answering the call for a standard reliability measure for coding data. *Commun. Methods Meas.* **2007**, *1*, 77–89. [CrossRef]

62. Proutskova, P.; Gruszczynski, M. *kripp.boot: Bootstrap Krippendorff's alpha Intercoder Reliability Statistic*, R package, version 1.0.0; 2017; Available online: https://github.com/MikeGruz/kripp.boot (accessed on 19 April 2019).

63. Kosmala, M.; Wiggins, A.; Swanson, A.; Simmons, B. Assessing data quality in citizen science. *Front. Ecol. Environ.* **2016**, *14*, 551–560. [CrossRef]

64. Rousselet, J.; Imbert, C.-E.; Dekri, A.; Garcia, J.; Goussard, F.; Vincent, B.; Denux, O.; Robinet, C.; Dorkeld, F.; Roques, A.; et al. Assessing species distribution using Google Street View: A pilot study with the pine processionary moth. *PLoS ONE* **2013**, *8*, e74918. [CrossRef]

65. Mooney, S.J.; Bader, M.D.M.; Lovasi, G.S.; Teitler, J.O.; Koenen, K.C.; Aiello, A.E.; Galea, S.; Goldmann, E.; Sheehan, D.M.; Rundle, A.G. Street audits to measure neighborhood disorder: Virtual or in-person? *Am. J. Epidemiol.* **2017**, *186*, 265–273. [CrossRef]

66. Baruch, A.; May, A.; Yu, D. The motivations, enablers and barriers for voluntary participation in an online crowdsourcing platform. *Comput. Hum. Behav.* **2016**, *64*, 923–931. [CrossRef]

67. Morschheuser, B.; Hamari, J.; Koivisto, J.; Maedche, A. Gamified crowdsourcing: Conceptualization, literature review, and future agenda. *Int. J. Hum. Comput. Stud.* **2017**, *106*, 26–43. [CrossRef]

68. Wiggins, A.; Newman, G.; Stevenson, R.; Crowston, K. Mechanisms for data quality and validation in citizen science. In Proceedings of the 2011 IEEE Seventh International Conference on e-Science Workshops, Stockholm, Sweden, 5–8 December 2011; 2011; pp. 14–19.

69. Gardiner, M.M.; Allee, L.L.; Brown, P.M.; Losey, J.E.; Roy, H.E.; Smyth, R.R. Lessons from lady beetles: Accuracy of monitoring data from US and UK citizen-science programs. *Front. Ecol. Environ.* **2012**, *10*, 471–476. [CrossRef]

70. Swanson, A.; Kosmala, M.; Lintott, C.; Packer, C. A generalized approach for producing, quantifying, and validating citizen science data from wildlife images. *Conserv. Biol.* **2016**, *30*, 520–531. [CrossRef]

forests

MDPI

Article

Do High-Voltage Power Transmission Lines Affect Forest Landscape and Vegetation Growth: Evidence from a Case for Southeastern of China

Xiang Li [1,2] and Yuying Lin [1,*]

[1]   College of Transportation and Civil Engineering, Fujian Agriculture and Forestry University, Fuzhou 350108, China; lixiang78162@163.com
[2]   Maintenance Branch Company of State Grid Fujian Electric Power Co. Ltd., Fuzhou 350013, China
*   Correspondence: linyuying2013@fafu.edu.cn; Tel.: +86-591-8378-9485

Received: 22 January 2019; Accepted: 12 February 2019; Published: 14 February 2019

check for
updates

Abstract: The rapid growth of the network of high-voltage power transmission lines (HVPTLs) is inevitably covering more forest domains. However, no direct quantitative measurements have been reported of the effects of HVPTLs on vegetation growth. Thus, the impacts of HVPTLs on vegetation growth are uncertain. Taking one of the areas with the highest forest coverage in China as an example, the upper reaches of the Minjiang River in Fujian Province, we quantitatively analyzed the effect of HVPTLs on forest landscape fragmentation and vegetation growth using Landsat imageries and forest inventory datasets. The results revealed that 0.9% of the forests became edge habitats assuming a 150 m depth-of-edge-influence by HVPTLs, and the forest plantations were the most exposed to HVPTLs among all the forest landscape types. Habitat fragmentation was the main consequence of HVPTL installation, which can be reduced by an increase in the patch density and a decrease in the mean patch area (MA), largest patch index (LPI), and effective mesh size (MESH). In all the landscape types, the forest plantation and the non-forest land were most affected by HVPTLs, with the LPI values decreasing by 44.1 and 20.8%, respectively. The values of MESH decreased by 44.2 and 32.2%, respectively. We found an obvious increasing trend in the values of the normalized difference vegetation index (NDVI) in 2016 and NDVI growth during the period of 2007 to 2016 with an increase in the distance from HVPTL. The turning points of stability were 60 to 90 meters for HVPTL corridors and 90 to 150 meters for HVPTL pylons, which indicates that the pylons have a much greater impact on NDVI and its growth than the lines. Our research provides valuable suggestions for vegetation protection, restoration, and wildfire management after the construction of HVPTLs.

Keywords: high-voltage power transmission lines; habitat fragmentation; landscape fragmentation; normalized difference vegetation index (NDVI)

---

1. Introduction

To meet the constantly growing electricity demands and to reduce electricity transmission losses, countries in Europe, North America, and Asia are attempting to develop high-voltage power transmission lines (HVPTLs) including pylons [1,2]. HVPTLs have become globally necessary components of power transmission infrastructure, covering 5.5 million km in 2014, with predictions that this will increase to 6.8 million km in 2020 [3]. Likewise, the development of HVPTL in China is expected to increase, to reduce electricity transmission energy losses, since long distances exist between power stations and end-users [4]. With the rapid growth of the electricity grid, HVPTLs will inevitably cover more complex environments, such as mountains and forests, compared to plains and cultivated lands [5]. The large-scale increase of this infrastructure necessitates the assessment of its degree of impact on vegetation. One of the typical environmental management strategies in the State Grid of China is to clear vegetation (e.g., grass and shrubs) along HVPTLs within a certain buffer zone (e.g., 50–100 m) every two to three years [6]. Forest fragmentation caused by the construction of HVPTLs has been reported in a previous study [7]. However, no direct quantitative measurements of the effects of HVPTL on vegetation growth have been reported. In order to minimize potentially negative effects on vegetation, as well as its associated impacts on wildfires and wildlife, further knowledge on the impact of HVPTL on vegetation is crucial.

Electricity systems are commonly composed of three sections: power generation plants, power transmission facilities, and consumers. Among the three sections, transmission facilities act as a bridge connecting the power produced in the plants and the end users. The elements of the transmission facilities include power lines and power pylons or towers [8]. The transmission of electricity covers large areas and can produce many negative impacts [9]. Consequently, the possible impacts of HVPTL have attracted considerable public attention [2,10], including risks to health and safety due to electric and magnetic fields [4,11], environmental risks from electromagnetic fields [10], biological effects of electromagnetic fields [4], visual and perception impacts [5,8], and property values [2]. For example, Tong et al. [11] studied the effects of the electric field of a 500 kV overhead transmission line on a building and found that high-voltage cables can generate intense electric fields and the fields vary across locations of buildings relative to their overhead cable. This provides a good reference for building design. Porsius et al. [10] investigated the nocebo responses to HVPTLs, suggesting an increasing tendency in the number of health occurrences after exposure to HVPTLs. Previous studies have reported that HVPTLs increase the risk of thunderstorm asthma and childhood leukemia [12] and has limited impacts on male reproductive capacity [4]. A monetary quantification evaluation for eliminating the decline in the aesthetic quality caused by overhead HVPTLs was conducted, revealing a regional variation in the willingness to pay across different landscape contexts in Italy [8]. Remote sensing technology has been applied to monitor electric transmission infrastructures. For example, Qin et al. [3] and Schmidt et al. [13] used Light Detection and Ranging (LiDAR) data and high-resolution aerial imagery to produce geospatial maps of electric power infrastructure, which are valuable for analysis, planning, and risk evaluation.

Many of the aforementioned studies focused on the impact of HVPTLs on humans; however, many other studies have examined the environmental and biological effects of HVPTLs [4]. For example, environmental effects indicate that corona ions generated by HVPTLs can change the surrounding electrical environment, thus increasing aerosol charge levels [14]. The concentration of charged particles was found to be more than two times greater than the mean background value [15]. In another study, calving site locations and area uses were compared during the calving period before, during, and after the construction of a power line in Norway. The findings indicated that power line disturbances do not cause avoidance effects for wild ungulates, whereas construction activities can induce a temporary reduction in area use [16]. Inspections of vegetation encroachment in the power line corridor have been conducted based on high spatial resolution hyperspectral imagery, satellite imagery, and LiDAR data [17,18] because vegetation growth plays a critical role in fire risk in power line corridors [19]. However, few studies have examined the vegetation effects of HVPTLs; hence, more research is needed to provide valuable recommendations for wildfire management and forest protection.

As airborne and aerospace remote sensors can quickly obtain a wide range of data from different altitudes, large scales, and multiple spectra, they have been extensively and successfully applied to meteorological observations, resource surveys, mapping, and military reconnaissance [20,21]. Satellite imagery has been successfully applied in HVPTL-related studies, such as high-resolution aerial imagery to classify electric power transmission lines [13] and LiDAR data to detect power line corridors [3]. The combination of high-resolution hyperspectral satellite images and LiDAR data was employed for vegetation management in HVPTL corridors [17], and satellite images have been combined with multimedia wireless sensor networks to improve the monitoring accuracy of vegetation. To the best of our knowledge, few studies have been published on the application of multi-spectral remote sensing in detecting the vegetation change response to HVPTLs. In multi-spectral remote sensing images, vegetation has the following characteristics due to the cellular structure of leaves: (1) High reflectance in the near-infrared (NIR) band and strong absorption in the red band due to chlorophyll; (2) by using an algorithm to divide the NIR by the red band (R), the vegetation area on the image has a relative height value and reaches a saturation value when the green biomass is high. Based on the characteristics of vegetation, the normalized difference vegetation index (NDVI) was introduced by Rouse et al. [22]. This is the current optimal indicator for identifying vegetation coverage and its growth status [23]. Due to the convenience and feasibility of detecting vegetation using remote sensing, we extracted the NDVI to quantify the vegetation change dynamics in HVPTL corridors.

In this study, the effects of HVPTL on the changes of vegetation and forest landscape fragmentation in the upper reaches of the Minjiang River of Fujian Province, China were observed based on a comprehensive analysis of multi-source data using remote sensing (RS) and geographic information systems (GIS). The objectives of this study were (1) to identify whether and to what extent the construction of HVPTLs, both line corridors and pylons, impacts vegetation growth in its edge areas, and (2) to quantify the impact of the construction of HVPTL on the forest landscape structure. Thus, our research provides valuable information for wildfire management in HVPTL corridors, and vegetation protection and restoration along the edge habitats beyond the HVPTL corridor.

2. Materials and Methods

2.1. Study Area

Sanming City was selected as the study site to assess changes in vegetation and forest landscape fragmentation before and after HVPTL construction. This study area (116°22′–118°39′ E, 25°30′–27°07′ N) is located in the western part of the Fujian Province in Southeastern China (Figure 1), in the upper reaches of the Minjiang River, which has the seventh highest annual runoff in China [24,25]. Sanming City was suitable for this study for two reasons: (1) It is located in the middle subtropical area and is unique for its climatic and terrain features, with high biological diversity and lush vegetation. The proportion of forest is higher than 80% in the region. (2) Due to HVPTL construction and development in recent years, the forest in this zone has experienced increasing disturbances (e.g., affected vegetation growth, forest fires, and reduced forest area). Maintaining a harmonious relationship between HVPTL extension and the changes in vegetation and forest landscape fragmentation is one of the key scientific issues for this region and other parts of the world.

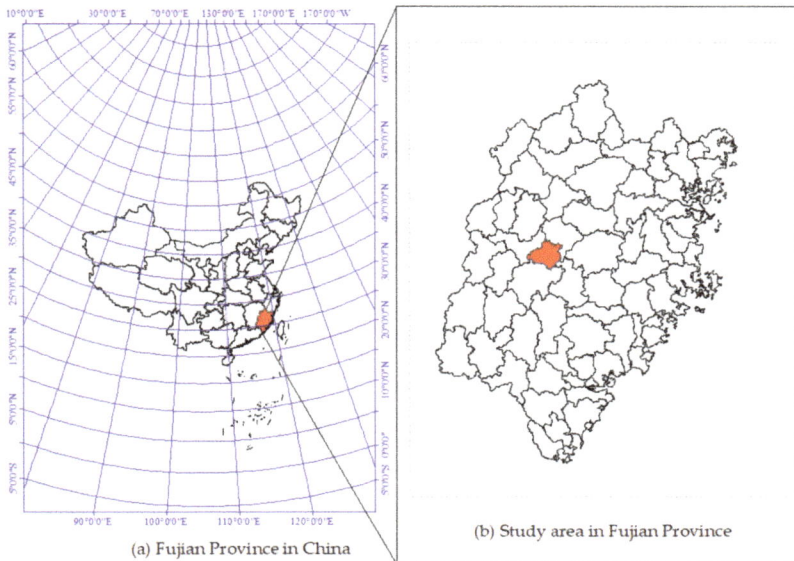

Figure 1. Location of the study area: (**a**) Fujian province in China and (**b**) study area in Fujian province.

2.2. Data Sources and Pre-Processing

The Forest Resources Inventory Database (FRID), NDVI, and HVPTL distribution map were used in this study. The FRID records forest characteristics and some growing factors at the forest patch level, such as tree species, tree age, tree height, slope, and altitude. The 2016 FRID, obtained from the local Forestry Bureau, was used to extract the spatial distribution of different forest vegetation types [24,26]. The NDVI data of the study area from 2007 and 2016 were extracted from Landsat remote sensing images. Landsat TM images from 2007 and Landsat OLI images from 2016 were acquired from the U.S. Geological Survey (USGS) (https://glovis.usgs.gov/) over a period of nine years. Data pre-processing included radiometric calibration, atmospheric calibration, and registration [27–29]. The HVPTL dataset of the study site was obtained from the Maintenance Branch Company of the State Grid Fujian Electric Power Co., Ltd. (Fuzhou, China).

2.3. Calculation of the NDVI

NDVI is currently the optimal indicator for identifying and assessing vegetation cover and growth status on different scales [30–32]. The effect of sensor degradation may be reduced by normalizing the spectral bands of the calculation of NDVI [32] as follows:

$$NDVI = (\rho_{\text{NIR}} - \rho_{\text{Red}}) / (\rho_{\text{NIR}} + \rho_{\text{Red}}) \tag{1}$$

where the ρ_{NIR} and ρ_{Red} are the planetary reflectance of infrared and near-infrared band in the TM and OLI sensors, respectively.

To evaluate the effects of HVPTLs on edge vegetation, 10 buffer zones, each with a width of 30 m, were created from the HVPTL outwards towards their edge vegetation. These buffers were then superimposed on the NDVI maps, and the average value of the vegetation index for each buffer was calculated using the spatial analysis tool of the GIS.

2.4. Landscape Classification

For a comprehensive description of the forest landscape structure and fragmentation of the study region, the FRID image was classified into five categories (Table 1, Figure 2): (1) semi-natural forest, (2) forest plantation, (3) bamboo forest, (4) other forest, and (5) non-forest land. A uniform spatial resolution (5 × 5 m) of the forest landscape map was used in this study.

Table 1. Classification of the landscape in the study area.

Landscape Classes	Description
Semi-natural forest	Forests that have re-grown after a timber harvest for a long enough period without human interference.
Forest plantation	Artificial mixed-species forest and artificial pure-species forest with artificially planted.
Bamboo forest	*Phyllostachys heterocycla, Dendrocalamopsis oldhami* and *D. latiflorus*, etc.
Other forest	Including shrubwood land, and sparse forest land, etc.
Non-forest land	Including construction land, cultivated land, water land, burned area, and barren land, etc.

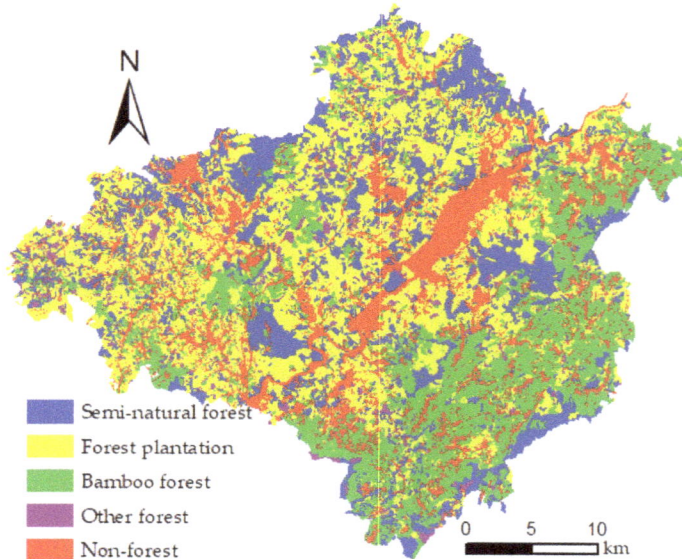

Figure 2. Spatial distribution of landscapes in study area.

2.5. Calculation of Landscape Metrics

To investigate the effects of the HVPTLs on forest landscape change, both landscape- and class-level indices were used to quantify the area, shape, and diversity of the patches and landscape (Table 2). For landscape-level measurements, the selected indices included patch density (PD), largest patch index (LPI), mean patch area (MA), area-weighted mean shape index (AWMSI), Shannon's diversity index (SHDI), Simpson's diversity index (SIDI), Shannon's evenness index (SHEI), and Simpson's evenness index (SIEI). For class-level measurements, the applied indices included PD, LPI, MA, and effective mesh size (MESH). These indices were calculated using the Fragstats 3.4 program to compare the landscape structure and fragmentation with and without HVPTL scenarios [33]. The calculation formula and corresponding descriptions of the above indices can be found in the Fragstats manual [33,34].

Table 2. The calculation formulas and descriptions for landscape metrics.

Landscape Metrics	Formula	Descriptions
Patches density (PD)	$PD = \frac{n_i}{A} \times 10000 \times 100$	Level: C/L; To describe the degree of fragmentation for a certain landscape type or a total landscape.
Largest patch index (LPI)	$LPI = \frac{max_{j=1}^{n}(a_{ij})}{A} \times 100$	Level: C/L; To provide a simple measure of dominance. It quantifies the percentage of total landscape area comprised by the largest patch.
Mean patch area (MA)	$MA = \frac{max_{j=1}^{n}(a_{ij})}{n_i} \times \frac{1}{10000}$	Level: C/L; To describe the degree of fragmentation for a certain landscape type or a total landscape.
Effective mesh size (MESH)	$MESH = \frac{\sum_{i=1}^{m}\sum_{j=1}^{n} a_{ij}^2}{A}$	Level: C; To indicate the probability of two points chosen randomly in a region will be connected.
Area-weighted mean shape index (AWMSI)	$AWMSI = \sum_{j=1}^{m}\left(0.25 p_{ij}/\sqrt{a_{ij}}\right)\cdot\left(a_{ij}/\sum_{j=1}^{m} a_{ij}\right)$	Level: L; To evaluate the shape complexity for the total landscape.
Shannon's diversity index (SHDI)	$SHDI = -\sum_{i=1}^{m} P_i log_2 P_i$	Level: L; To estimate the level of landscape diversity. SHDI is somewhat more sensitive to rare patch types than SIDI.
Simpson's diversity index (SIDI)	$SIDI = 1 - \sum_{i=1}^{m} P_i^2$	Level: L; It is another popular diversity measure. Compared with SHDI, the value of Simpson's index represents the probability that any two pixels would be different patch types.
Shannon's evenness index (SHEI)	$SHEI = \frac{-\sum_{i=1}^{m} P_i log_2 P_i}{log_2 m}$	Level: L; To describe the even distribution among patches. Evenness is the complement of dominance of certain patch.
Simpson's evenness index (SIEI)	$SIEI = \frac{1 - \sum_{i=1}^{m} P_i^2}{1 - \left(\frac{1}{m}\right)}$	Level: L; Similar as SHEI.

Note: L: Landscape-level; C: Class-level; n_i: patch number of landscape type i; A: total landscape area; a_{ij}: area of patch ij; p_{ij}: perimeter of patch ij; m: number of patch types (classes); P_i: area proportion of patch type (class) i to total landscape.

3. Results and Discussion

3.1. Edge Habitat Impacts of the High-Voltage Power Transmission Lines

Previous studies suggested that the edge effect distances along roads vary considerably from several meters to several hundred meters [35,36]. For comparison, we analyzed the effect of HVPTLs on edge habitat using 10 buffers with 30 m intervals (Figure 3). The introduction of the HVPTLs to the forest landscape resulted in 0.9% (1052 ha) of the vegetation as a whole becoming HVPTL edge habitat, assuming a 150 m depth-of-edge-influence, and 1.8% (2119 ha) assuming a 300 m depth-of-edge-influence. Then, we calculated the proportion of each forest type in each buffer along the edge habitat areas. This showed that the proportion of forests gradually increased with increasing distance from HVPTLs for semi-natural forest (from 23.1 to 26.1%) and the non-forest land (from 3.7 to 6.5%). The proportion of the other forest types declined from 5.0 to 2.4% as the distance increased from HVPTLs. The proportion of the forest plantation type decreased notably at first, then increased slightly with increasing distance from the HVPTL. Bamboo forest showed an opposite trend to the plantation, with an initial slight increase and then decreasing obviously with increased distance. As shown in

Figure 3, among all the landscapes, the proportions of the forest plantation were the highest, accounting for more than 50% of all the 10 HVPTL depths-of-edge. The proportions of the semi-natural forest ranked second among all the landscapes, accounting for 20 to 30% of all the 10 depths-of-edge of the HVPTLs. The proportions of non-forest and other forest types were relatively low, accounting for 2 to 7% of all 10 depths-of-edge of the HVPTLs. This effect showed that the forest plantations and semi-natural forests were the habitat types that were most affected by HVPTLs in the study area. The reason for this is that the study area is dominated by forests, accounting for more than 80% of the total land area, and the plantations (i.e., forest plantations) are the main components of the forest landscape (Figure 2).

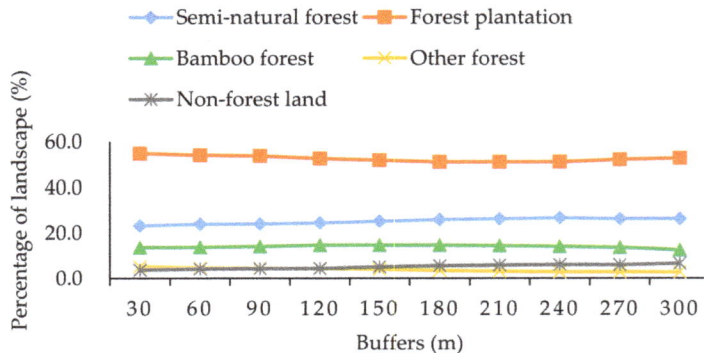

Figure 3. Percentage of each landscape type at the edge of high-voltage power transmission lines (HVPTLs) in 2016.

Using GIS to calculate the area of various forest landscapes (Figure 2), the proportions of forest plantations, semi-natural forests, bamboo forests, other forests, and non-forest lands were 36.6, 19.6, 23.5, 2.83, and 17.5%, respectively. The proportion of forest plantation in the HVPTL corridor was significantly higher than in the entire study area. This indicates that the forest plantation was the most affected landscape type during and after the construction of HVPTL, whereas bamboo forests and non-forest lands were less affected by the construction of HVPTL. This is in line with a previous finding, which indicated that power lines inevitably cross more forests with the rapid development of the network grid [3]. This, in turn, increases forest fragmentation to a certain degree [7]. This is discussed in the next section.

3.2. Effects of HVPTL on Forest Landscape Fragmentation

Combining the HVPTLs with the forest landscape map of 2016 showed the effects of HVPTLs on the structure of the forest landscape by introducing landscape metrics at both the landscape and class levels (Figures 4 and 5). The introduction of HVPTLs to the 2016 landscape increased the number and proportion of patches by 1.610 and 1.9%, respectively, and decreased the largest patch index by 2.297 and 20.8%, respectively (Figure 4). Mean patch area decreased by 0.271 and 2.0%, and the area-weighted mean shape index decreased from 12.135 to 9.896, with a decreasing ratio of 18.5% (Figure 4). These measures indicate that HVPTL causes an overall degree of fragmentation in the forest landscape. However, little effect of HVPTLs was observed on the diversity (i.e., SHDI and SIDI) and dominance (i.e., SHEI and SIEI) of the landscape, which means there is no significant difference in the overall heterogeneity of the landscapes with and without HVPTL.

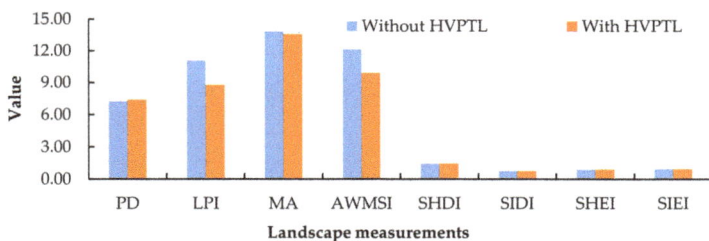

Figure 4. Landscape-level measurements with and without HVPTLs.

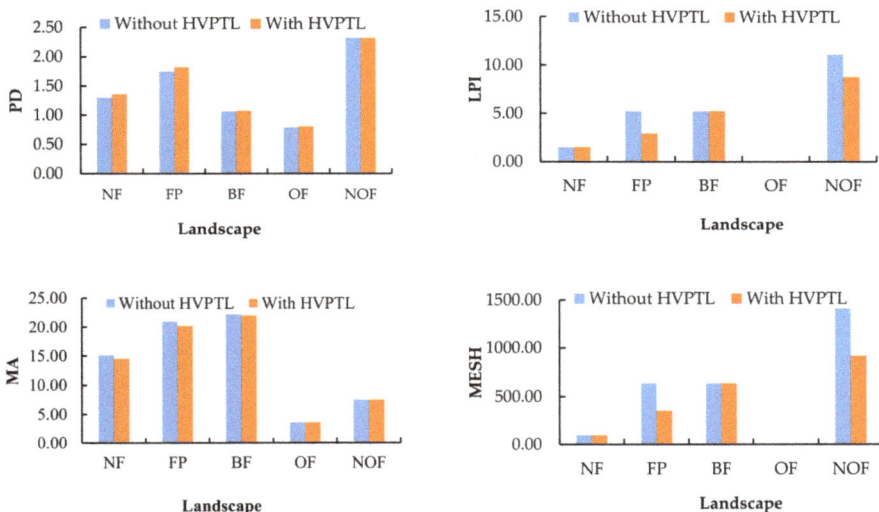

Figure 5. Landscape patch-class level measurements with and without HVPTLs. NF: semi-natural forest; FP: forest plantation; BF: bamboo forest; OF: other forest; NOF: non-forest.

Concerning the HVPTL effects of fragmentation on each landscape type, Figure 5 shows that the class-level measurements indicated a clear pattern of fragmentation caused by HVPTLs in each landscape. This is illustrated by the increase in the PD and decrease in the MA, LPI, and MESH. In all landscape types, forest plantation and non-forest land were most affected by HVPTL construction, with the values of LPI decreasing by 44.1 and 20.8%, respectively, and the values of MESH decreasing by 44.2 and 32.2%, respectively.

Even though the overall landscape diversity has not been strongly affected by HVPTL construction, subtler habitat fragmentations may have a significant impact on the conservation of sensitive interior forest species [37]. For example, the implications of decreasing MA and MESH, combined with the aforementioned area-of-edge-influence, manifest largely in the reduction of the amount of interior habitat in forest plantations available for sensitive interior species. The corridors created by the HVPTLs can increase human disturbances of the sensitive interior species. Notably, the most affected habitat type—forest plantation—has a relatively low species richness in the study area.

3.3. Effects of HVPTL on Vegetation Growth Dynamics

To observe the effects of HVPTLs (both lines and pylons) on NDVI, we delimited 10 buffer zones from high voltage lines and pylons to 30 m away (Figures 6–9).

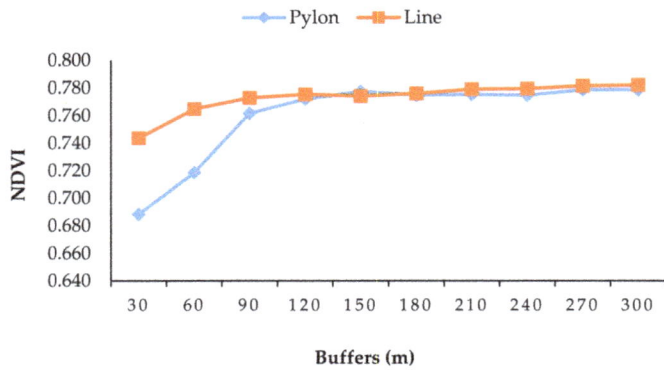

Figure 6. Normalized difference vegetation index (NDVI) value varies with buffer gradient based on data from 2016.

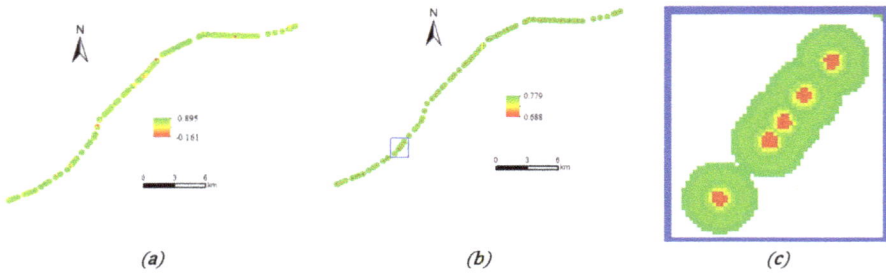

Figure 7. Spatial distribution of NDVI at the edge of HVPTL pylons in 2016: (**a**) The distribution of the original value of NDVI; (**b**) overall distribution of the average value of NDVI in each buffer zone; and (**c**) the average value of NDVI in each buffer zone of the extracted location.

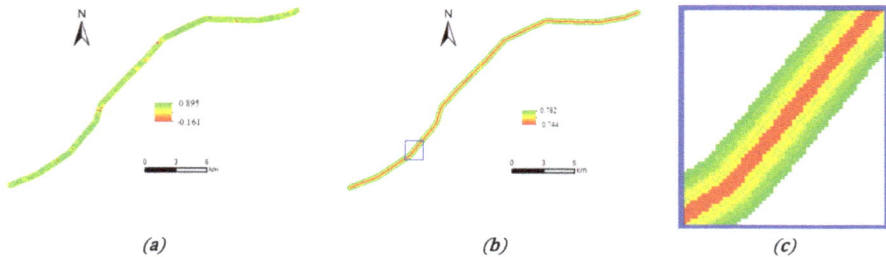

Figure 8. Spatial distribution of NDVI at the edge of HVPTL lines in 2016: (**a**) The distribution of the original value of NDVI; (**b**) the overall distribution of the average value of NDVI in each buffer zone; and (**c**) the average value of NDVI in each buffer zone of the extracted location.

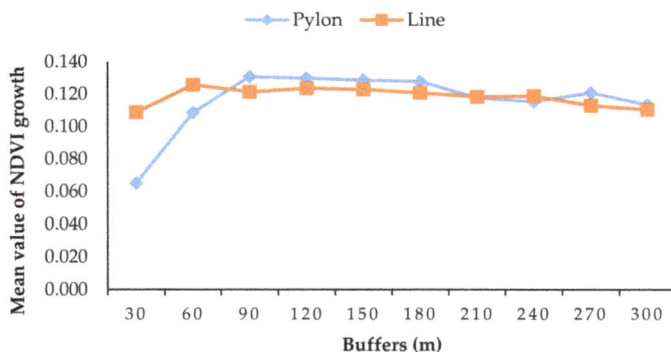

Figure 9. NDVI growth during 2007–2016 varies with the buffer width.

Figures 6–8 show a similar trend between the pylons' and the lines' effects, with an obviously increasing trend in the NDVI value with increasing distance from the HVPTLs. The difference between the two curves is that the NDVI value in the line reached a smooth level 90 m away, whereas the NDVI value for the pylons reached a smooth level 150 m away. The NDVI values for the lines were significantly higher than those for the pylons before reaching a plateau. Then, their values were similar after reaching the peak. Simultaneously, we analyzed the NDVI growth in the corridors (i.e., buffer zones) of HVPTLs, which revealed that NDVI showed different degrees of increasing trends in different buffer zones (Figure 9). However, the trend in NDVI growth varied in different buffers, with the NDVI growing slower near HVPTLs. NDVI growth tended to be stable 60 m from the HVPTL lines, whereas they became stable at 90 m from HVPTL pylons. These results reveal that pylons have a much greater impact on NDVI and its value than the high-voltage lines, regardless of the magnitude or range of the influence.

The reasons for the above effects of HVPTLs on NDVI are as follows: (1) The construction of HVPTLs is the primary driving factor that induces a temporary reduction in NDVI and its growth in the corridors. According to the regulations of the State Grid of China on environmental management [6], the vegetation along the HVPTL corridors (within 50 to 100 m) is usually cleared every two or three years. This was verified using the results of this study, which show that the NDVI within the 50–90 m buffer zone is significantly lower than in the farther regions, and the NDVI growth within the 60–90 m buffer zone is significantly lower than in the farther buffers. (2) In contrast with the irregular edges created by natural disturbances (e.g., fire and windthrow)—where there is progressive vegetation recovery, especially in the subtropical region (i.e., this study area) with sufficient hydrothermal conditions—HVPTL corridors tend to exist long-term and suffer from frequent disturbances, e.g., vegetation clearance every two to three years. (3) The forest edge experiences microclimatic changes, including increased evaporation, increased temperature, solar radiation enhancement, and soil moisture reduction [38]. (4) Some previous findings indicated that the strong electric fields generated from power lines impact human health [11], whereas others have revealed that the electric field may not be a disturbance for wild ungulates [16]. Though there are many applications of satellite imagery for vegetation growth near power lines [18], whether high-voltage magnetic fields and their electric fields affect NDVI is still unclear and requires further research.

4. Conclusions

To minimize the potentially negative effects on vegetation and closely associated wildfire, further knowledge on the impact of HVPTLs on vegetation and forests is required. To this end, taking one of the areas with the highest forest coverage in China as a case (the upper reaches of the Minjiang River in Fujian Province), we quantitatively analyzed the effect of HVPTLs on forest landscape patterns and vegetation growth using Landsat images and forest inventory datasets.

Forests **2019**, *10*, 162

The results revealed that 1.8% of the vegetation as a whole becomes edge habitat, assuming a 300 m depth-of-edge-influence by HVPTLs. Forest plantations were substantially more exposed to HVPTLs compared with semi-natural forests, bamboo forests, and other forests.

Habitat fragmentation was the main HVPTL effect; this was highlighted by the increase in patch density and decrease in MA, LPI, and MESH. In all the landscape types, forest plantations and non-forest land were most affected by HVPTLs, with the values of LPI decreasing by 44.1 and 20.8%, respectively, and the values of MESH decreasing by 44.2 and 32.2%, respectively.

Finally, we found that NDVI values increased with increasing distance from HVPTLs in 2016 and that NDVI growth increased from 2007 to 2016 with increasing distance from HVPTLs. Concerning the NDVI in 2016, the NDVI value for HVPTL plateaued at 90 m from the HVPTLs, whereas the NDVI value for the pylons plateaued at 150 m from HVPTLs. From 2007 to 2016, NDVI growth tended to stabilize 60 m from the HVPTL and 90 m from the pylons. This indicates that the pylons have a much greater impact on NDVI and its growth than the lines.

We provide strong evidence that HVPTLs occupy a considerable proportion of forest land, which can have a potentially detrimental impact not only on wildlife habitats but also on wildfire management. Quantitative remote sensing methods combined with ordinary GIS software enables analyses to be easily and quickly replicated. These analyses could provide decision support for vegetation protection, restoration, and wildfire management after the construction of HVPTLs.

Author Contributions: Conceptualization, Y.-Y.L.; Methodology, X.L.; Software, X.L.; Validation, X.L. and Y.-Y.L.; Formal Analysis, X.L. and Y.-Y.L.; Investigation, X.L.; Resources, X.L. and Y.-Y.L.; Data Curation, X.L. and Y.-Y.L.; Writing-Original Draft Preparation, X.L.; Writing-Review & Editing, Y.-Y.L.; Visualization, X.L. and Y.-Y.L.; Supervision, Y.-Y.L.

Funding: This research was funded by the National Natural Science Foundation of China under grant number 41201100 and the Natural Science Foundation of Fujian under grant number 2015J01606.

Acknowledgments: We are very grateful to all those who provided the raw data used for our research, especially to the Maintenance Branch Company of State Grid Fujian Electric Power Co.,Ltd for providing us with the HVPTL dataset, and the Forestry Bureau of Sanming City in Fujian Province providing us with the forest inventory datasets. This study was funded by the Natural Science Foundation of Fujian (No. 2015J01606), to which we are also very grateful. And many thanks for the support provided by the National Natural Science Foundation of China (No.41201100).

Conflicts of Interest: The authors declare no conflict of interest.

References

1. Bamigbola, O.; Ali, M.; Oke, M. Mathematical modeling of electric power flow and the minimization of power losses on transmission lines. *Appl. Math. Comput.* **2014**, *241*, 214–221. [CrossRef]
2. Cain, N.L.; Nelson, H.T. What drives opposition to high-voltage transmission lines? *Land Use Policy* **2013**, *33*, 204–213. [CrossRef]
3. Qin, X.; Wu, G.; Ye, X.; Huang, L.; Lei, J. A Novel Method to Reconstruct Overhead High-Voltage Power Lines Using Cable Inspection Robot LiDAR Data. *Remote Sens.* **2017**, *9*. [CrossRef]
4. Wu, S.; Di, G.; Li, Z. Does static electric field from ultra-high voltage direct-current transmission lines affect male reproductive capacity? Evidence from a laboratory study on male mice. *Environ. Sci. Pollut. Res.* **2017**, *24*, 18025–18034. [CrossRef] [PubMed]
5. Soini, K.; Pouta, E.; Salmiovirta, M.; Uusitalo, M.; Kivinen, T. Local residents' perceptions of energy landscape: the case of transmission lines. *Land Use Policy* **2011**, *28*, 294–305. [CrossRef]
6. Xu, K.; Zhang, X.; Chen, Z.; Wu, W.; Li, T. Risk assessment for wildfire occurrence in high-voltage power line corridors by using remote-sensing techniques: A case study in Hubei Province, China. *Int. J. Remote Sens.* **2016**, *37*, 4818–4837. [CrossRef]
7. Luken, J.O.; Hinton, A.C.; Baker, D.G. Forest edges associated with power-line corridors and implications for corridor siting. *Landsc. Urban Plan.* **1991**, *20*, 315–324. [CrossRef]
8. Tempesta, T.; Vecchiato, D.; Girardi, P. The landscape benefits of the burial of high voltage power lines: A study in rural areas of Italy. *Landsc. Urban Plan.* **2014**, *126*, 53–64. [CrossRef]

9.   Doukas, H.; Karakosta, C.; Flamos, A.; Psarras, J. Electric power transmission: An overview of associated burdens. *Int. J. Energy Res.* **2010**, *35*, 979–988. [CrossRef]

10. Porsius, J.T.; Claassen, L.; Woudenberg, F.; Smid, T.; Timmermans, D.R. Nocebo responses to high-voltage power lines: Evidence from a prospective field study. *Sci. Total Environ.* **2016**, *543*, 432–438. [CrossRef] [PubMed]

11. Tong, Z.; Dong, Z.; Ashton, T. Analysis of electric field influence on buildings under high-voltage transmission lines. *IET Sci. Meas. Technol.* **2016**, *10*, 253–258. [CrossRef]

12. Redmayne, M. A proposed explanation for thunderstorm asthma and leukemia risk near high-voltage power lines: A supported hypothesis. *Electromagn. Biol. Med.* **2018**, *37*, 57–65. [CrossRef] [PubMed]

13. Schmidt, E.H.; Bhaduri, B.L.; Nagle, N.; Ralston, B.A. Supervised Classification of Electric Power Transmission Line Nominal Voltage from High-Resolution Aerial Imagery. *GISci. Remote Sens.* **2018**, *55*, 1–20. [CrossRef]

14. Wright, M.D.; Buckley, A.J.; Matthews, J.C.; Shallcross, D.E.; Henshaw, D.L. Air ion mobility spectra and concentrations upwind and downwind of overhead AC high voltage power lines. *Atmos. Environ.* **2014**, *95*, 296–304. [CrossRef]

15. Jayaratne, E.; Ling, X.; Morawska, L. Comparison of charged nanoparticle concentrations near busy roads and overhead high-voltage power lines. *Sci. Total Environ.* **2015**, *526*, 14–18. [CrossRef] [PubMed]

16. Colman, J.E.; Tsegaye, D.; Flydal, K.; Rivrud, I.M.; Reimers, E.; Eftestøl, S. High-voltage power lines near wild reindeer calving areas. *Eur. J. Wildl. Res.* **2015**, *61*, 881–893. [CrossRef]

17. Frank, M.; Pan, Z.; Raber, B.; Lenart, C. Vegetation management of utility corridors using high-resolution hyperspectral imaging and LiDAR. Proceedings of 2010 2nd Workshop on Hyperspectral Image and Signal Processing: Evolution in Remote Sensing, Reykjavik, Iceland, 14–16 June 2010; IEEE: Piscataway, NJ, USA, 2010.

18. Ahmad, J.; Malik, A.S.; Xia, L. Effective techniques for vegetation monitoring of transmission lines right-of-ways. Proceedings of 2011 IEEE International Conference on Imaging Systems and Techniques, Penang, Malaysia, 17–18 May 2011; IEEE: Piscataway, NJ, USA, 2011.

19. Zhang, H.; Han, X.; Dai, S. Fire Occurrence Probability Mapping of Northeast China With Binary Logistic Regression Model. *IEEE J-STARS* **2013**, *6*, 121–127. [CrossRef]

20. Hakkenberg, C.; Zhu, K.; Peet, R.; Song, C. Mapping multi-scale vascular plant richness in a forest landscape with integrated LiDAR and hyperspectral remote-sensing. *Ecology* **2018**, *99*, 474–487. [CrossRef] [PubMed]

21. Lees, K.; Quaife, T.; Artz, R.; Khomik, M.; Clark, J. Potential for using remote sensing to estimate carbon fluxes across northern peatlands—A review. *Sci. Total Environ.* **2018**, *615*, 857–874. [CrossRef] [PubMed]

22. Rouse, J.W.; Haas, R.H.; Scheel, J.A.; Deering, D.W. Monitoring Vegetation Systems in the Great Plains with ERTS. In Proceedings of the 3rd Earth Resource Technology Satellite (ERTS) Symposium, Washington, DC, USA, 10–14 December 1973.

23. Carlson, T.N.; Ripley, D.A. On the relation between NDVI, fractional vegetation cover, and leaf area index. *Remote Sens. Environ.* **1997**, *62*, 241–252. [CrossRef]

24. Lin, Y.; Hu, X.; Zheng, X.; Hou, X.; Zhang, Z.; Zhou, X.; Qiu, R.; Lin, J. Spatial variations in the relationships between road network and landscape ecological risks in the highest forest coverage region of China. *Ecol. Indic.* **2019**, *96*, 392–403. [CrossRef]

25. Hu, X.; Wu, C.; Hong, W.; Qiu, R.; Li, J.; Hong, T. Forest cover change and its drivers in the upstream area of the Minjiang River, China. *Ecol. Indic.* **2014**, *46*, 121–128. [CrossRef]

26. Guan, J.; Zhou, H.; Deng, L.; Zhang, J.; Du, S. Forest biomass carbon storage from multiple inventories over the past 30 years in Gansu Province, China: implications from the age structure of major forest types. *J. For. Res.* **2015**, *26*, 887–896. [CrossRef]

27. Chander, G.; Markham, B.L.; Helder, D.L. Summary of current radiometric calibration coefficients for Landsat MSS, TM, ETM+, and EO-1 ALI sensors. *Remote Sens. Environ.* **2009**, *113*, 893–903. [CrossRef]

28. Charvz, P.S., Jr. Image-based atmospheric corrections—Revisited and revised. *Photogramm. Eng. Rem. S.* **1996**, *62*, 1025–1036.

29. Xu, H.; Zhang, T. Assessment of consistency in forest-dominated vegetation observations between ASTER and Landsat ETM plus images in subtropical coastal areas of southeastern China. *Agric. For. Meteorol.* **2013**, *168*, 1. [CrossRef]

30. Chu, H.; Venevsky, S.; Wu, C.; Wang, M. NDVI-based vegetation dynamics and its response to climate changes at Amur-Heilongjiang River Basin from 1982 to 2015. *Sci. Total Environ.* **2019**, *650*, 2051–2062. [CrossRef]

31. Joiner, J.; Yoshida, Y.; Anderson, M.; Holmes, T.; Hain, C.; Reichle, R.; Koster, R.; Middleton, E.; Zeng, F.-W. Global relationships among traditional reflectance vegetation indices (NDVI and NDII), evapotranspiration (ET), and soil moisture variability on weekly timescales. *Remote Sens. Environ.* **2018**, *219*, 339–352. [CrossRef]

32. Hu, X.; Xu, H. A new remote sensing index for assessing the spatial heterogeneity in urban ecological quality: A case from Fuzhou City, China. *Ecol. Indic.* **2018**, *89*, 11–21. [CrossRef]

33. Liu, S.; Dong, Y.; Deng, L.; Liu, Q.; Zhao, H.; Dong, S. Forest fragmentation and landscape connectivity change associated with road network extension and city expansion: A case study in the Lancang River Valley. *Ecol. Indic.* **2014**, *36*, 160–168. [CrossRef]

34. Liu, S.; Deng, L.; Chen, L.; Li, J.; Dong, S.; Zhao, H. Landscape network approach to assess ecological impacts of road projects on biological conservation. *Chin. Geogr. Sci.* **2014**, *24*, 5–14. [CrossRef]

35. McGarigal, K.; Romme, W.H.; Crist, M.; Roworth, E. Cumulative effects of roads and logging on landscape structure in the San Juan Mountains, Colorado (USA). *Landsc. Ecol.* **2001**, *16*, 327–349. [CrossRef]

36. Forman, R.T.T. Estimate of the Area Affected Ecologically by the Road System in the United States. *Conserv. Biol.* **2000**, *14*, 31–35. [CrossRef]

37. Hejl, S.J. The importance of landscape patterns to bird diversity: a perspective from the northern Rocky Mountains. *Northwest Environ. J.* **1992**, *8*, 119–137.

38. Reed, R.A.; Johnson-Barnard, J.; Baker, W.L. Contribution of Roads to Forest Fragmentation in the Rocky Mountains. *Conserv. Biol.* **1996**, *10*, 1098–1106. [CrossRef]

![forests logo] *forests*

MDPI

Article

Mapping Forest Canopy Height in Mountainous Areas Using ZiYuan-3 Stereo Images and Landsat Data

Mingbo Liu [1,2], Chunxiang Cao [1,*], Yongfeng Dang [3] and Xiliang Ni [1]

[1] Institute of Remote Sensing and Digital Earth, Chinese Academy of Sciences, Beijing 100101, China; liumb@radi.ac.cn (M.L.); nixl@radi.ac.cn (X.N.)
[2] University of Chinese Academy of Sciences, Beijing 100049, China
[3] Academy of Forest Inventory and Planning, State Forestry Administration, Beijing 100714, China; dangyf_9205@163.com
*   Correspondence: caocx@radi.ac.cn; Tel.: +86-010-6483-6205

Received: 11 December 2018; Accepted: 28 January 2019; Published: 29 January 2019

check for updates

Abstract: Forest canopy height is an important parameter for studying biodiversity and the carbon cycle. A variety of techniques for mapping forest height using remote sensing data have been successfully developed in recent years. However, the demands for forest height mapping in practical applications are often not met, due to the lack of corresponding remote sensing data. In such cases, it would be useful to exploit the latest, cheaper datasets and combine them with free datasets for the mapping of forest canopy height. In this study, we proposed a method that combined ZiYuan-3 (ZY-3) stereo images, Shuttle Radar Topography Mission global 1 arc second data (SRTMGL1), and Landsat 8 Operational Land Imager (OLI) surface reflectance data. The method consisted of three procedures: First, we extracted a digital surface model (DSM) from the ZY-3, using photogrammetry methods and subtracted the SRTMGL1 to obtain a crude canopy height model (CHM). Second, we refined the crude CHM and correlated it with the topographically corrected Landsat 8 surface reflectance data, the vegetation indices, and the forest types through a Random Forest model. Third, we extrapolated the model to the entire study area covered by the Landsat data, and obtained a wall-to-wall forest canopy height product with 30 m × 30 m spatial resolution. The performance of the model was evaluated by the Random Forest's out-of-bag estimation, which yielded a coefficient of determination (R^2) of 0.53 and a root mean square error (RMSE) of 3.28 m. We validated the predicted forest canopy height using the mean forest height measured in the field survey plots. The validation result showed an R^2 of 0.62 and a RMSE of 2.64 m.

Keywords: forest canopy height; ZiYuan-3 stereo images; SRTMGL1; digital photogrammetry; Landsat 8; mountainous areas

1. Introduction

Forest canopy height refers to the mean height of the top surface of a forest's canopy. It is useful in studying biodiversity and the carbon cycle. Many studies have mapped forest canopy height using remote sensing data [1–9]. The majority of these are based on the light detection and ranging (LiDAR), synthetic aperture radar (SAR) or digital photogrammetry (DP) techniques [5,8,10].

Researchers have demonstrated the feasibility of airborne laser scanning (ALS) in mapping forest canopy height under various conditions [10–12]. The Geoscience Laser Altimeter System (GLAS) onboard NASA's Ice, Cloud, and land Elevation Satellite (ICESat) produced the most commonly used spaceborne LiDAR data, which can also be used to map forest height [13]. Researchers also mapped forest height by combining the LiDAR and auxiliary data, such as the meteorological data and the

Moderate Resolution Imaging Spectroradiometer (MODIS) data [2,3,14–16]. The SAR is well known for its high temporal resolution, because it is less affected by weather and illumination conditions [17]. The SAR-based techniques for forest canopy height mapping include radargrammetry, interferometry SAR (InSAR), and polarimetric interferometric SAR (PolInSAR). The radargrammetry is based on SAR stereo images [5,18,19]. The InSAR is based on the phase differences between two complex SAR images [6,20]. The PolInSAR includes phase difference methods and the model-based methods, which rely on the complex coherent coefficient model [7,21]. In these methods, the TerraSAR-X (X-band), Radarsat-2 (C-band), and ALOS-2 (L-band) are commonly used SAR data sources. It is expected that the techniques for mapping forest canopy height using the LiDAR and SAR will be further developed when data from the ICESat-2 [22], the GEDI [23], the Biomass (P-band) (due for launch in 2021) [24], and the Tandem-L (L-band) (due for launch in 2022) [25] missions become available. The airborne DP systems have a large field of view, high cruising altitude, fast data acquisition speed, and easy flight planning [26]. The spaceborne DP systems can perform continuous and repetitive observations. Several spaceborne systems, such as the WorldView series, provide sub-meter resolution stereo images from which high-quality digital surface models (DSM) can be extracted [27]. Since the DSM represents the surface of the earth, including vegetation, buildings and other man-made features, extracting canopy height requires a digital terrain model (DTM) that provides the bare earth elevation (another term that needs to be distinguished here is the digital elevation model or DEM, which is a generic term that can refer to either a DSM or DTM). In recent years, some studies have mapped forest canopy height using airborne or spaceborne ultra-high spatial resolution (finer than 1 m × 1 m) stereo images and the ALS-derived DTM data [9,28–32].

Existing techniques provide a variety of means for mapping forest height. The ALS is probably the most promising and mature technique. However, the demands for forest canopy height mapping in practical applications are often not met. Because the data needed in the existing techniques (e.g., LiDAR or SAR data) may not be available when it comes to a given study area. The lack of data demonstrates the necessity of making full use of the free datasets, exploiting the latest, cheaper datasets, and combining them with models for forest height mapping. In this study, we used the ZiYuan-3 (ZY-3) stereo images to derive the DSM and the Shuttle Radar Topography Mission global 1 arc second data (SRTMGL1) as the DTM. We correlated the extracted canopy height model (CHM) with the Landsat data through a regression model and obtained a wall-to-wall forest canopy height product after extrapolation.

2. Study Area

Our study area is located in the Yangtze River Basin, upstream from the Three Gorges Dam in Hubei Province, Central China (30°13′40″–31°34′32″ N, 110°4′54″–111°39′11″ E) (Figure 1). This is a transitional zone between the mountains and plains with a total area of 11,339.70 km². The region is ecologically fragile and dominated by forest ecosystems. It is an important area for ecological protection and biodiversity conservation. The region's ecological significance is the main reason we chose it as our study area. Due to the completion of the Three Gorges Reservoir, the changes in the ecological environment in this region have received considerable attention in the last decade. We expect that the forest canopy height product obtained in this study will serve as a reference for related future studies.

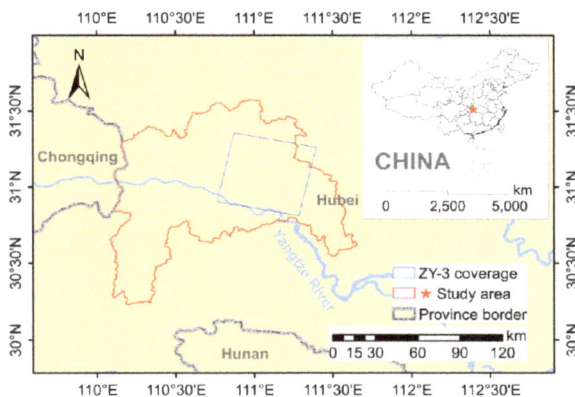

Figure 1. Location of the study area.

The study area we selected is mountainous and the terrain is complex. The altitude ranges from 3 m to 2995 m (Figure 2). The ground slope is between 0° and 77°, with an average of 23.40°. The dominant communities include *Pinus massoniana* Lamb. forest, *Cunninghamia lanceolate* (Lamb.) Hook. forest, *Quercus variabilis* Blume forest, *Cyclobalanopsis glauca* (Thunb.) Oerst. forest and *Pinus armandii* Franch. forest. The main vegetation types are coniferous forest, evergreen broad-leaved forest, evergreen and deciduous broad-leaved mixed forest, deciduous broad-leaved forest, coniferous and broad-leaved mixed forest, and shrub [33].

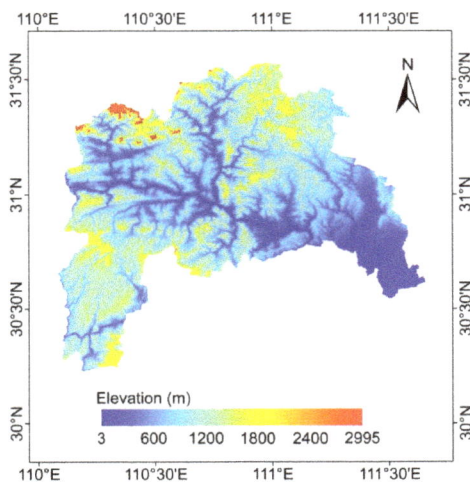

Figure 2. Elevation (SRTMGL1) of the study area.

3. Data

3.1. ZY-3 Data

The ZY-3 satellite, launched on 9 January 2012, is China's first civilian stereo mapping satellite. It carries one multi-spectral sensor, and three panchromatic sensors pointing forward, backward, and nadir directions, respectively. All three panchromatic sensors have spectral ranges of 0.50 μm–0.80 μm. The forward and backward sensors have resolutions of 3.5 m × 3.5 m, while the nadir sensor has a resolution of 2.1 m × 2.1 m. The angle between the nadir sensor and the other two sensors is 23.5°. The multi-spectral sensor consists of four channels: Blue (0.45 µm–0.52 µm), green (0.52 µm–0.59 µm), red (0.63 µm–0.69 µm), and near infrared (0.77 µm–0.89 µm), with resolutions of 5.8 m × 5.8 m. The positioning error of the ZY-3 is less than 4 m and the height measurement error is less than 3 m if the ground control points (GCPs) are available [34–36]. The ground swath is about 50 km × 50 km and the digitalizing bit is 10 bits [37]. The ZY-3 data used in this study were acquired on 8 October 2014, during the leaf-on season. We ordered the data from the website of the China Centre for Resources Satellite Data and Application (CRESDA, http://www.cresda.com/CN/). As the forward view image contained more mountain shadows, we only used the backward and nadir view panchromatic images to extract the DSM. We used the multi-spectral image to classify the land-cover and extract forest areas.

3.2. SRTMGL1 Data

The Shuttle Radar Topography Mission (SRTM) data were collected from 11 February 2000 to 22 February 2000, with a radar interferometry system [38–42]. The SRTMGL1 data for China have been available since July 2015 [43]. The spatial resolution of this data is about 30 m × 30 m in the study area. Elevation values are stored as 16-bit signed integers in the HGT files. The vertical accuracy of SRTMGL1 in Hubei Province ranges from 1.9 m to 18.6 m, with higher accuracy in flatter areas. In this study, we relied on the SRTMGL1 to determine the elevation of the GCPs and used it as the DTM to calculate the slope and aspect. We downloaded SRTMGL1 Version 3 product from NASA's website (https://earthdata.nasa.gov/). This is a void-filled version, with fill value information included in the ancillary NUM files.

We also used the SRTMGL1 as the DTM to extract the crude CHM in this study. The penetration capability of the SRTM C-radar was one of the main reasons for this choice. There are also several other near-global terrain data available [43]. The ASTER Global Digital Elevation Model (GDEM) and ALOS World 3D-30 m DEM (AW3D30) are based on optical stereo images from multiple periods and cannot penetrate the forest, in principle. The terrain data produced by the TerraSAR-X add-on for Digital Elevation Measurements (TanDEM-X) mission are also DSMs because the X-band radar beam can hardly penetrate the canopy [44,45]. It was reported that the phase center of the SRTM C-radar is located above the ground in many cases, but specific analysis is required. Forest height, structure, and density all influence penetration depth. Studies have shown that the SRTM radar phase center can still reach the land surface if the forest height is relatively low (e.g., less than 30 m, as was the case in our study area) [39,46]. In addition, the SRTM radar beam should penetrate deeper in leaf-off seasons (i.e., February 11 to February 22, 2000), because during these times, the bare land surface is visible even in optical images. If the phase center is close to the land surface, or the penetration depth increases as the height of the forest increases, using the SRTMGL1 as a DTM remains worthwhile.

3.3. Landsat 8 OLI Surface Reflectance Data

The Landsat 8 Operational Land Imager (OLI) surface reflectance product is derived from the Landsat 8 Level-1 precision and terrain corrected product (L1TP), using the Landsat 8 Surface Reflectance Code (LaSRC) algorithm [47]. The acquisition date of the Landsat 8 data used in this study is 15 September 2013. We used them for the land-cover classification, principal component analysis (PCA), image segmentation, and to calculate the vegetation indices as well as to predict the forest canopy height. We also collected the Landsat 8 data acquired in the non-growing season to help with the classification of forest types. The acquisition date of the non-growing season data is 21 January 2014. We obtained these data from NASA's website (https://earthdata.nasa.gov/). The images acquired in 2013 were contaminated by a few clouds and haze on the southwestern and eastern edges of the study area (Figure 14). The effects of haze were mitigated by the LaSRC algorithm. The clouds were masked using the Level-2 Pixel Quality Assurance band in the Landsat 8 surface reflectance product. The cloud flags in the Level-2 Pixel Quality Assurance band were derived from the Function of Mask (Fmask) algorithm, with the high-confidence cloud pixels dilated [47,48]. A few remaining cloud pixels were eliminated after the land-cover classification. The images acquired in 2014 were cloudless but contained more mountain shadows. The shadows in the images were also masked after the land-cover classification.

3.4. Field Data

We obtained field survey data from the forestry administration. The survey plots were set near the coordinate grid nodes on topographic maps and were located in intact forests. The plots are square or rectangular and cover an area of 0.667 hectares. Field measurements were undertaken in the leaf-off season of 2013. The diameter at breast height (DBH) of all the trees with a DBH greater than 5 cm in each plot was recorded. Then, three to five trees, with a DBH close to the average, were selected and their heights were derived from angle and distance measurements. Angles were measured using a mechanical clinometer. Distances were measured using a laser range finder. The average height of the selected trees was recorded as the mean height of the forest. The measured mean forest height was used to validate the predicted forest canopy height of this study.

4. Methods

Figure 3 displays a flow chart of the methods we used in this study. These methods can be divided into four sections. First, we derived a DSM from the ZY-3 stereo images and subtracted the SRTMGL1 to obtain a crude CHM. We then applied a series of refining measures to the crude CHM, including the masking of non-forest areas using land-cover classification results obtained from the ZY-3 multi-spectral image. Second, we correlated the refined CHM with the topographically corrected Landsat 8 surface reflectance data, the vegetation indices, and the forest types using a Random Forest regression model. Third, we extrapolated the model to the entire study area covered by the Landsat 8 data. Finally, we compared our predicted forest canopy height with the mean forest height measured in the field survey plots.

4.1. Extraction and Refinement of Crude CHM

We extracted the DSM from the ZY-3 stereo images using the OrthoEngine module of the PCI Geomatica software (PCI Geomatics Enterprises, Inc., Canada). The backward view image was automatically re-sampled to the same resolution as the nadir view image. The module uses the polynomial coefficients, GCPs, and tie points to compute a math model that relates the rows and columns of the matching pixels with ground coordinates and elevations. The polynomial coefficients are provided in the rational polynomial coefficients (RPC) files distributed with the ZY-3 data. We set forty-nine GCPs manually in the vegetation-free areas referencing the ZY-3 multi-spectral image, ESRI's online World Imagery, and Google Earth (Figure 4). The elevation of the GCPs was derived from the SRTMGL1 to minimize the elevation differences between the ZY-3 DSM and SRTMGL1. We collected a total of 208 tie points interactively in the nadir and backward view images. We set the sampling distance of the output DSM to 2 m × 2 m, to avoid the loss of accuracy.

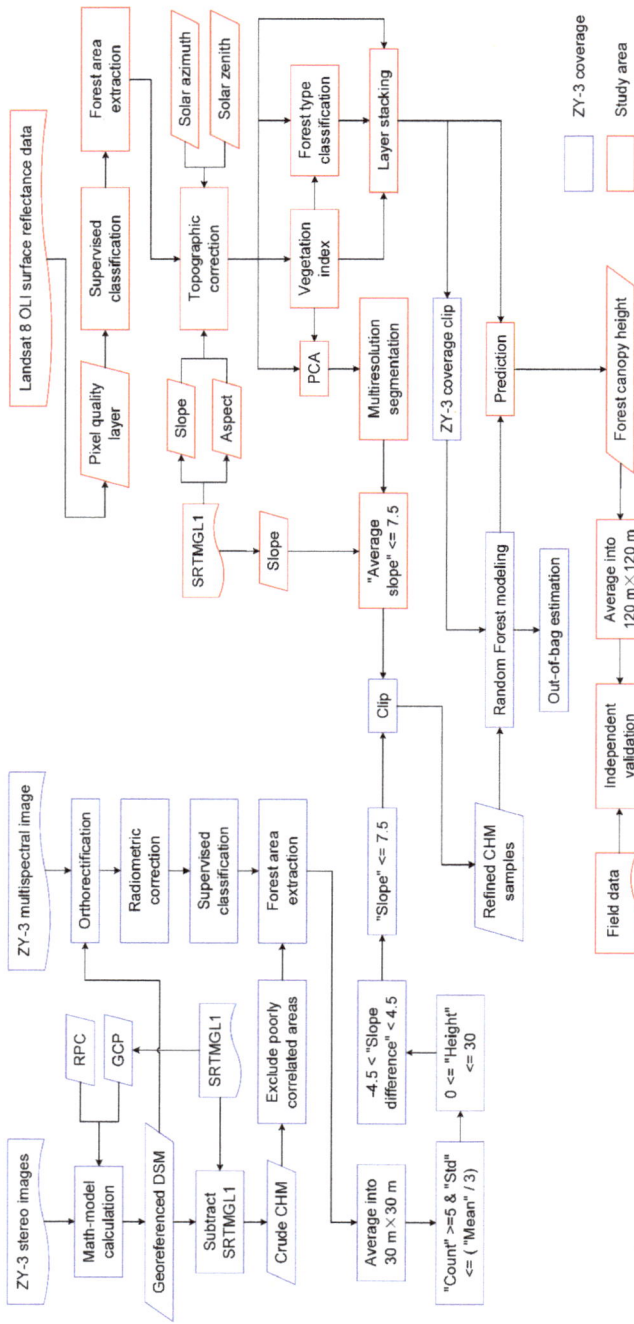

Figure 3. Flow chart of methods. The blue box indicates that the data and operations were based on the entire study area. The red box indicates that the data and operations were based on the ZY-3 coverage. In the flow chart, RPC stands for Rational Polynomial Coefficients files, GCP stands for Ground Control Points, and PCA stands for Principal Component Analysis.

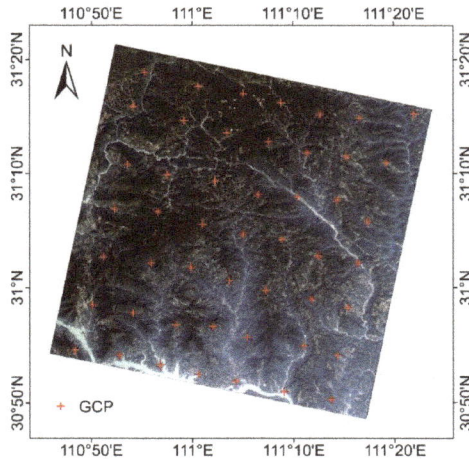

Figure 4. The GCPs and ZY-3 multi-spectral image. We set 49 GCPs manually in the vegetation-free areas and derived the elevation of the GCPs from the SRTMGL1.

The extracted ZY-3 DSM can only represent the elevation of the upper surface of the forest canopy. We used bi-linear interpolated 2 m × 2 m resolution SRTMGL1 data as the DTM and subtracted it from the 2 m × 2 m resolution ZY-3 DSM to obtain a crude CHM (Figure 5). We had removed the filled values from the SRTMGL1 in advance according to the auxiliary NUM files.

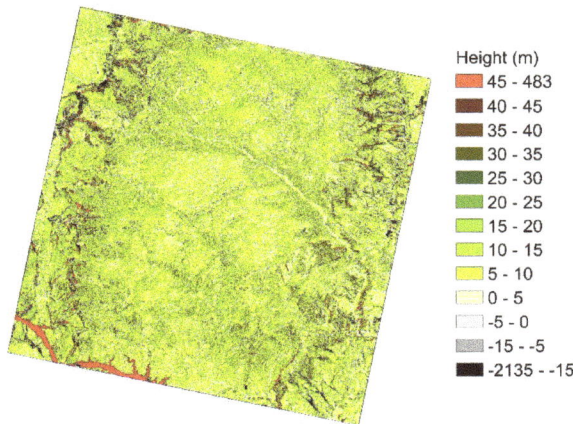

Figure 5. The crude canopy height model (CHM) extracted from the ZY-3 DSM and SRTMGL1.

In the crude CHM, forests with canopy heights larger than 15 m are mostly concentrated in the north and mid-west. This corresponds to the locations of protected areas and scenic spots. Subgraphs (b) and (f), (c) and (g), and (d) and (h) in Figure 6 show the vertical information of different surface objects, such as forest patches and new houses.

Rising water levels after the 2009 completion of the Three Gorges Dam are responsible for the extracted CHM above the Yangtze River. We also found numerous erroneous values in rugged places, especially on the left and right edges of the ZY-3 coverage, which could come from shadows and steep terrain. As Figure 6e shows, both the CHM values above 45 m (the red areas) and below −15 m (the black areas), were erroneous. They correspond to the areas with shadows and steep terrain in

Figure 6a. In addition, resolution differences could also cause errors, which would be mentioned later. The erroneous values in the crude CHM highlighted the need for data refining.

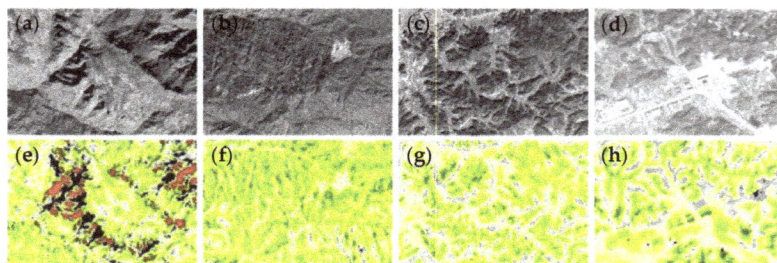

Figure 6. The ZY-3 nadir view images and the corresponding crude CHMs. The crude CHM's legend is the same as Figure 5's. Subgraphs (**a**) and (**e**) show the impact of shadows and steep terrain, with the positive error labelled in red and the negative error labelled in black. (**b**) and (**f**), (**c**) and (**g**), and (**d**) and (**h**) show the vertical information of different surface objects.

We refined the crude CHM through a series of measures. First, the areas with poor correlation between stereo images were excluded. We achieved this by masking pixels with scores equal to zero. The score layer was generated during the DSM extraction process. Pixels with scores equal to zero are places where the matching between the stereo images failed, and their values were filled based on the surrounding values. Since they were not calculated from the geometric model, they needed to be excluded.

Second, we ortho-rectified the multi-spectral image, acquired simultaneously with the stereo images using the ZY-3 DSM. We then radiometrically corrected it, based on the calibration parameters downloaded from the CRESDA (http://www.cresda.com/CN/). Next, we used the maximum likelihood supervised classification method to classify the multi-spectral image into five categories. Training samples were selected in Google Earth. The number of samples for forest, water body, shadow, farmland, and bare surface were 80, 20, 50, 95, and 90, respectively. After the classification, only the CHM pixels in forest areas were preserved. Third, we averaged the 2 m × 2 m resolution CHM pixels into 30 m × 30 m to align with the pixels of the Landsat. In the process of averaging, we removed the areas with pixel counts less than 5, or standard deviations larger than 1/3 of the mean values in each 30 m × 30 m cell, which correspond to the regions with poor uniformity. These regions are susceptible to the averaging operation. A total of 61.85% of the averaged CHM pixels were excluded in this way. Fourth, we eliminated 2.80% of CHM pixels, with values less than 0 m or larger than 30 m. This threshold was determined based on the information provided by the forestry department. Few forests in this region have mean canopy heights greater than 30 m. Therefore, canopy heights greater than 30 m are more likely to represent abnormal values.

Fifth, the ZY-3 DSM has a finer resolution and contains more topographical details such as hills and depressions than the SRTMGL1. This means that when we subtracted the SRTMGL1 from the ZY-3 DSM, these topographical details would be transferred to the crude CHM and mix with the forest canopy heights. To solve this problem, we used the slope difference layer between the ZY-3 DSM and the SRTMGL1 as a mask. In order to rule out the undulations on the canopy surface, we averaged the ZY-3 DSM to the same resolution as the SRTMGL1 before calculating the slopes and their differences. As Figure 7 shows, we only preserved the CHM pixels with slope differences less than 4.5°. We relied on visual inspections referencing the stereoscopic display of the terrain data to select the slope difference threshold. We found that the areas being masked increased sharply as the threshold decreased. The 4.5° threshold was a compromise option to minimize the interferences from terrain, while retaining enough samples. The masked parts correspond to the rugged areas where the CHM is abnormal. A total of 17.85% of the CHM pixels were excluded in this step.

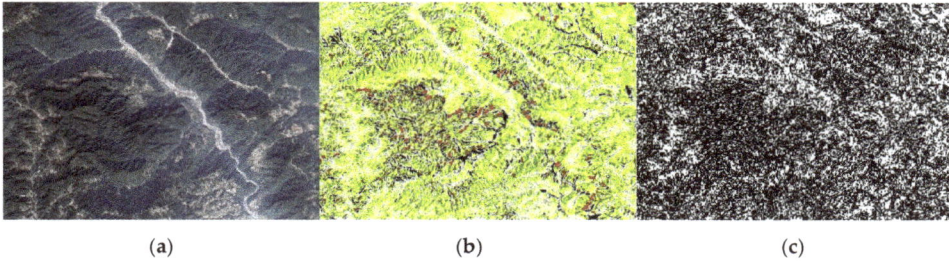

Figure 7. The slope difference mask. Subgraph (**a**) shows the ZY-3 multi-spectral image of the demonstration area. Subgraph (**b**) shows the crude CHM that was contaminated by terrain fluctuations (the legend is the same as Figure 5's). Subgraph (**c**) shows the slope difference mask. We only preserved the areas with slope differences less than 4.5° (displayed in white). The 4.5° threshold was a compromise option to minimize the interferences from terrain, while retaining enough samples.

Sixth, to further reduce the impact of the terrain, the CHM pixels with slopes larger than 7.5° were removed, accounting for 16.96% of all CHM pixels. This threshold was also a compromise option to suppress terrain interference while maintaining sufficient CHM pixels. In addition, we referenced the filtering methods of the GLAS data in mountainous areas. In those cases, the slope threshold was set to 5° to 7° [3,49].

4.2. Preparation of Landsat Data

We used the Level-2 Pixel Quality Assurance band in the Landsat 8 surface reflectance product, in order to assess the per-pixel quality. In each scene, only the pixels flagged as clear land pixels were preserved. We then classified the clear land pixels using the maximum likelihood supervised classification method. Training samples were selected in Google Earth. The number of samples for forest, water body, shadow, cloud, farmland, and bare surface were 200, 45, 120, 30, 52, and 150, respectively. We only used the pixels classified as forest in the subsequent processing and modeling.

We applied the C-correction [50] to correct the effects of illumination and terrain on surface reflectance. It can be calculated as follows:

$$\cos(i) = \cos(sz) \times \cos(sl) + \sin(sz) \times \sin(sl) \times \cos(az - as) \tag{1}$$

$$\rho_{\lambda,t} = b_\lambda + m_\lambda \times \cos(i) \tag{2}$$

$$c_\lambda = b_\lambda / m_\lambda \tag{3}$$

$$\rho_{\lambda,n} = \rho_{\lambda,t} \times (\cos(sz) + c_\lambda) / (\cos(i) + c_\lambda) \tag{4}$$

where $\cos(i)$ is the cosine of the solar incidence angle, sz is solar zenith angle, sl is slope, az is solar azimuth angle, as is aspect, $\rho_{\lambda,t}$ is the spectral reflectance influenced by topography, b_λ and m_λ are intercept and slope of the linear regression model between $\rho_{\lambda,t}$ and $\cos(i)$. c_λ is the empirical parameter calculated separately for each spectral band λ. $\rho_{\lambda,n}$ is the topographically corrected spectral reflectance. The slope and aspect data used in Equation (1) were derived from the SRTMGL1. The solar azimuth and zenith data used in Equation (1) and Equation (2) were provided in the Landsat surface reflectance product.

Next, we calculated the following four vegetation indices: The normalized difference vegetation index (NDVI), the enhanced vegetation index (EVI), the soil adjusted vegetation index (SAVI), and the modified soil adjusted vegetation index (MSAVI).

We classified the pixels of the forest area into different forest types. The classification of forest types was based on four vegetation index layers derived from the topographically corrected Landsat 8 non-growing season scenes (acquired on January 21, 2014). Again, in this case, we used the maximum likelihood supervised classification method. A total of 100 training samples of deciduous forest, and 100 training samples of evergreen forest, were selected by referencing multi-temporal high spatial resolution images from Google Earth. We also performed a forest type classification based on the reflectance bands and NDVI of the topographically corrected Landsat 8 growing season scenes (acquired on September 15, 2013), using the same set of training samples. This was used to fill the void areas caused by shadows in the non-growing season classification result.

We performed PCA on seven reflectance bands and four vegetation indices of the Landsat data. The first and second principal components contributed 93.22% of the accumulative eigenvalue. They were stacked with the forest type layer to serve as the input data of the multi-resolution segmentation. We used the eCognition Developer software (Trimble Germany GmbH, Germany) to implement the multi-resolution segmentation. Then, we calculated the average slope in each segment based on the slope layer derived from the SRTMGL1. The segments with average slopes of no more than 7.5° were selected to clip the CHM layer. This threshold was also determined by experiments to suppress terrain interferences, while retaining sufficient CHM pixels. A total of 0.5% CHM pixels were excluded in this way. Finally, after excluding a total of 99.96% of crude CHM pixels, we obtained the refined CHM samples for use in subsequent modeling.

4.3. Random Forest Modeling and Extrapolation

Statistical modeling and extrapolation were used in this study for two reasons. First, because we discarded most of the crude CHM pixels during the refining process, the refined CHM samples might be very sparse. Thus, we needed to use the statistical model to obtain a wall-to-wall forest canopy height map. Second, the coverage of the study area is much larger than that of the ZY-3. The use of a statistical model is more economically efficient. It will reduce the dependence on the abundance of the ZY-3 data and increase the update frequency of the forest canopy height product.

Random Forest [51] is an ensemble of unpruned decision trees, induced from bootstrap samples. Each tree is grown with randomized subsets of variables and functions by recursively partitioning the dependent variable into less varying subsets. A prediction is made by averaging the predictions of the ensemble. It is robust to noise and has good generalization ability. We employed the Random Forest algorithm to correlate the refined CHM with the topographically corrected Landsat 8 surface reflectance data, the vegetation indices, and the forest types. The SAVI and MSAVI were not used in the regression model as they showed strong correlations with the NDVI and EVI. All the 10 predictor layers (seven reflectance layers, two vegetation index layers, and a forest type layer) were clipped to the ZY-3 coverage. The Random Forest's built-in out-of-bag estimation was used to evaluate the performance of the model. There are two important parameters can be used to tune the model. The mtry is the number of variables tried at each split. The ntree is the number of decision trees. We determined these parameters by experiments. The values that produced the best out-of-bag estimation result were selected. The mtry was set to 4, and the ntree was set to 200 after the experiments. Variable importance was evaluated by the percentage increase in the mean squared error (%IncMSE) and the increase in node purity (IncNodePurity). The %IncMSE is the increase of mean squared error after randomly permuting the values of the variable, averaged over all trees and normalized by the standard deviation of the differences. The IncNodePurity is calculated by adding up the decreases of residual sum of squares when the variable splits nodes, calculated over all trees.

We used the random Forest package in R environment (R Development Core Team, http://www.R-project.org) to implement the Random Forest algorithm. We extrapolated the regression model to the entire study area to obtain a wall-to-wall forest canopy height product. The predicted forest canopy height was independently validated using the mean forest height measured in the field survey plots.

5. Results and Discussion

5.1. Refined CHM Samples

Figure 8 shows the refined CHM samples. Most of the samples were located in the relatively flat areas between the mountains. Some were at the tops of the mountains. The number of samples was 1,020. Their mean was 9.55 m, close to the reference mean forest height of 8.59 m provided by the forestry department for this area. The spatial resolution of the refined CHM samples was 30 m × 30 m.

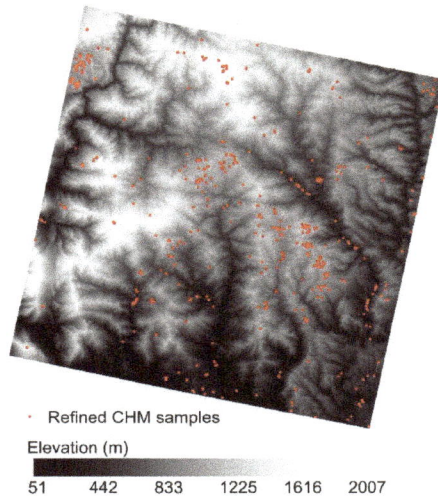

· Refined CHM samples

Elevation (m)

| 51 | 442 | 833 | 1225 | 1616 | 2007 |

Figure 8. Refined CHM samples on the elevation layer.

5.2. Forest Type Classification

Figure 9 shows the forest type classification result. To evaluate the classification accuracy, we randomly set 500 points in the study area and used manual interpretation in Google Earth to determine the reference categories for each point.

The number of points used for assessing the accuracy of classification was determined based on a balance between what was statistically sound and what was practically attainable. Congalton proposed an empirical minimum number of points of 50–100 for each category. He also pointed out that this number should be adjusted according to the relative importance and the inherent variability of each category [52]. In our case, a total of 321 points fell into the deciduous forest areas, 143 points fell into the evergreen forest areas, and 36 points fell into the non-forest areas. Non-forest areas (accounting for 7.27% of the total area) got relatively less points. But this was acceptable, since they were mainly composed of waters with small variabilities and were less important relative to forests. At the same time, the workload of manually interpreting 500 points remained attainable.

The confusion matrix showed producer accuracy values of 0.62, 0.94, and 0.78 for non-forest area, deciduous forest, and evergreen forest, respectively. The user accuracy values were 0.89, 0.84, and 0.87, respectively. The total accuracy was 0.86 and the kappa coefficient was 0.73.

It was found that the deciduous forests were widely distributed throughout the study area. They covered a total area of 7,275.02 km^2 and had an average distribution-altitude of 923 m. The evergreen forests showed a clustered distribution. They covered a total area of 3,240.54 km^2 and had an average distribution-altitude of 917 m.

Figure 9. Classification result of forest types.

5.3. PCA and Multi-resolution Segmentation

Figure 10 shows a piece of the composited layer of the first principal component (PC1), the second principal component (PC2), and the forest type. The red channel represents PC1, the green channel represents PC2, and the blue channel represents forest type.

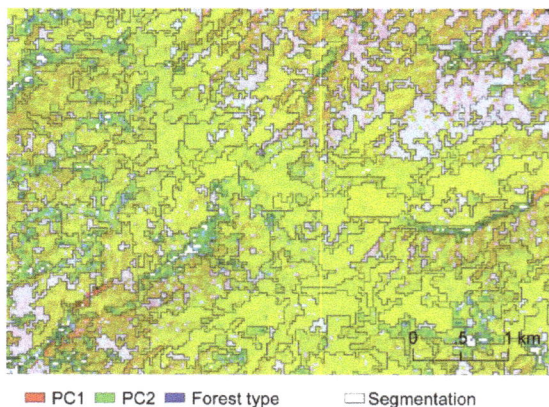

Figure 10. A piece of the composited layer of the first principal component (PC1), the second principal component (PC2), and the forest type. We performed multi-resolution segmentation, based on the composited layer.

We performed multi-resolution segmentation based on the composited layer. Segmentation parameters were set interactively to ensure that different forest types were separated and the segment size was about 1 km. Finally, we obtained 357,628 segments with a mean area of 3.17 hectares.

5.4. 30 m × 30 m Resolution Forest Canopy Height Mapping

As Figure 11 shows, band5 (Near infrared band) and band3 (Green band) were the most important variables in the Random Forest model. Band5 had the highest %IncMSE (18.84%) and band3 had the most significant impact on the IncNodePurity. Band1 (Coastal/Aerosol band) and band4 (Red band) also had relatively high levels of %IncMSE, and IncNodePurity, respectively.

The importance of bands 3 (0.53 μm–0.59 μm) and 5 (0.85 μm–0.88 μm) have a bio-physical basis. Band3 is related to chlorophyll. A more intense reflectance in this band usually corresponds to younger forests and therefore, corresponds to a lower canopy height. Band5 is associated with biomass, and biomass is highly correlated with forest canopy height. Therefore, we suggest that they deserve closer attention in future research. If studies use multi-spectral data, other than Landsat, they should make the existence of these two bands a prerequisite.

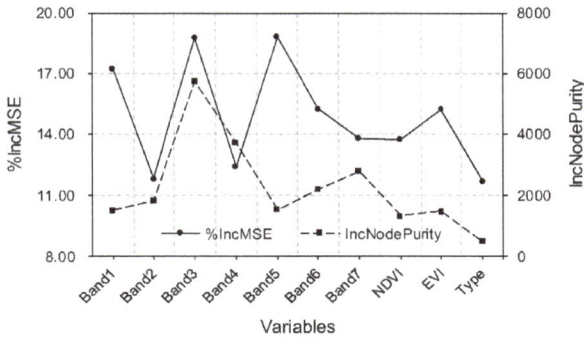

Figure 11. Variable importance of the Random Forest model.

We performed a piecewise averaging of all samples to reveal the relationships between predictor variables and refined CHM (Appendix A). It was found that, with the increase of the refined CHM, the reflectance of all bands showed a downward trend, and the NDVI and EVI showed upward trends. Evergreen forests usually accounted for larger proportions in the groups with higher refined CHMs. We used out-of-bag estimation to evaluate the Random Forest regression model. The canopy height of each sample was predicted, based on about 1/3 of the regression trees. The estimation showed a coefficient of determination (R^2) of 0.53 and a root mean square error (RMSE) of 3.28 m (Figure 12).

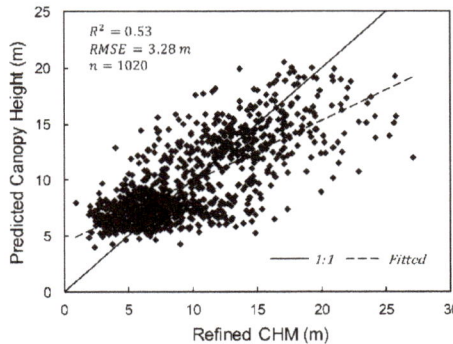

Figure 12. Out-of-bag estimation of the Random Forest model.

Figure 13 shows the predicted forest canopy height, which was related to the forest type to some extent. For example, two areas, one northeast of the Three Gorges Dam and the other east of the study area, had relatively large canopy heights and both of them were evergreen coniferous forest areas.

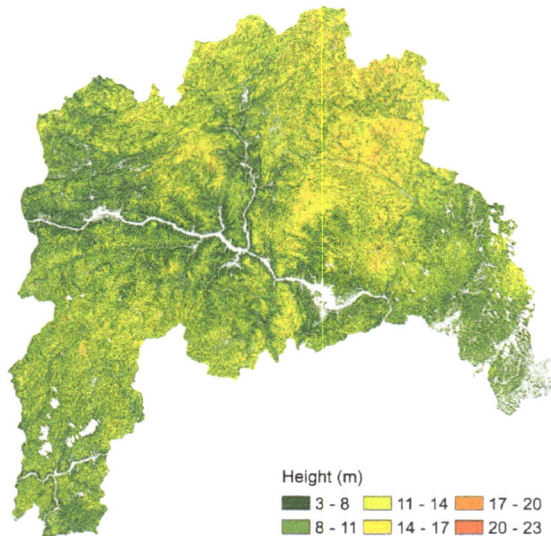

Height (m)
■ 3 - 8 ▢ 11 - 14 ■ 17 - 20
■ 8 - 11 ▢ 14 - 17 ■ 20 - 23

Figure 13. Forest canopy height predicted by the Random Forest model.

The predicted forest canopy height also responded to altitude. In high mountain areas, forest canopy heights gradually decreased as the altitude rose, forming sunken patches centered on the mountain peaks. Some alpine regions in the northwest corner (red areas in Figure 2) of the study area are covered with snow for months every year and the forest canopy heights in these areas were also low.

The study area has long been a region with frequent human activity. Except for some planted forests on the hillsides, orchards and tea trees dominate the areas below the altitude of 800 m. The predicted forest canopy height confirmed this. As Figure 13 shows, except for some planted forests in the east, the forests located below the altitude of 800 m had lower canopy heights than other forests. The spatial distributions of these forests were also fragmented. In contrast, forests in protected areas and scenic spots had higher canopy heights and more continuous distributions.

Our methods produced three kinds of canopy heights: (1) The crude CHM had a resolution of 2 m × 2 m and represented the heights of different positions on the canopy surface. (2) During the refining process, we averaged the crude CHM in 30 m × 30 m cells. Thus, the refined CHM had a resolution of 30 m × 30 m and could be regarded as a mean canopy height. (3) We used the refined CHM as the target value in the Random Forest model, thus the predicted canopy height also represented the mean canopy height.

5.5. Independent Validation of the Predicted Forest Canopy Height

Figure 14 shows the spatial distribution of the 20 field survey plots on the Landsat image. They are distributed between 390 m and 1520 m above sea level, and their slopes range from 8° to 42°. The age of the forests surveyed is between 8 years and 48 years. The measured mean forest heights range from 5.2 m to 15.2 m, with an average of 8.79 m. All the survey plots used for validation are occupied by coniferous and broad-leaved mixed forests. Considering that the forest type classification in this study was rough and the forest type did not play a significant role in the model (Figure 11), the survey plots occupied by other forest types were not used.

Before validation, we averaged the predicted forest canopy height in a 4 × 4 pixel window around the GPS position of each survey plot to reduce the effects of GPS positioning errors and the residuals of topographic correction. The validation result (Figure 15) showed an R^2 of 0.62 and a RMSE of 4.66 m. An offset of 6.5 m was observed. After compensating the offset, the RMSE dropped to 2.64 m.

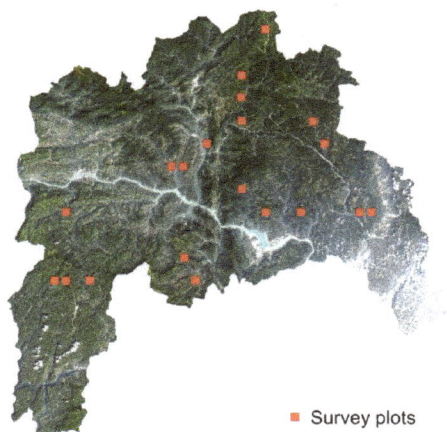

Figure 14. Spatial distribution of the survey plots.

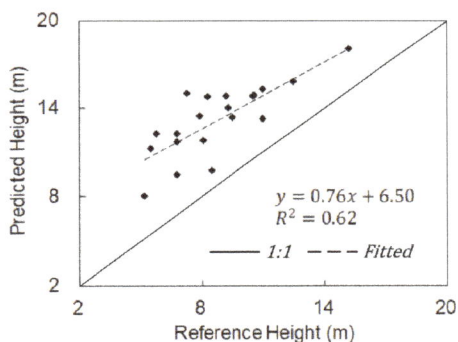

Figure 15. Independent validation of the predicted forest canopy height.

Two factors might be responsible for the offset in the validation. First, we used 49 GCPs set in vegetation-free areas when extracting the ZY-3 DSM. Most of the GCPs were located on artificial surfaces, such as roads and farmlands in valleys. Since the SRTMGL1 has a spatial resolution of 30 m × 30 m, the extracted elevation of GCPs might be affected by the surrounding hillsides and be higher than the actual elevation. This might shift the ZY-3 DSM upwards, causing an over-estimation of the CHM and, in turn, an over-estimation of the predicted forest canopy height. The second factor, perhaps a more influential factor, might be the topographic correction. An insufficient topographic correction would result in lower reflectance in shady slopes. This would also lead to an over-estimation of the predicted forest canopy height, since the reflectance had negative correlation with canopy height, as shown in Figure A1. Further studies may require more data, with higher spatial resolution and vertical accuracy. In the future, we plan to apply this method to areas covered by ALS data for a more comprehensive assessment. The ALS data can also be used to calibrate the possible offset of the predicted forest canopy height.

5.6. Limitations and Future Outlook

In the refining of the crude CHM, we excluded pixels from areas of moderate or high relief, which is a limitation of this study. However, it was necessary for exploring the correlation between the CHM and predictor variables, since the forest height information was overwhelmed by erroneous values in many places. We believe that to remove this limitation, a more complicated and ingenious method is needed to refine the crude CHM, which is the direction of our future research. In addition, two assumptions guided the extrapolation of the model. First, we assumed that the topographic correction effectively eliminated the effects of terrain and illumination on the reflectance. This meant that the same object should have the same Landsat reflectance, regardless of the slope, aspect, and light source direction. Second, we assumed that the forests with the same type and Landsat reflectance would have the same canopy heights, no matter where they were. However, we had observed many small-scale residuals from the topographic correction in the rugged areas. We also greatly simplified the classification of forest types. These are issues that need to be addressed in future research.

The resolution difference between the ZY-3 DSM and SRTMGL1 had a significant impact on the CHM. The ZY-3 DSM had a finer resolution and contained more topographical details than the SRTMGL1. Although we interpolated the SRTMGL1 to 2 m × 2 m resolution before subtracting it from the ZY-3 DSM, small-scale elevation changes on the ZY-3 DSM would still be transferred to the crude CHM and mix with the forest canopy heights. The resolution of the DTM also limited the use of other multi-spectral images with higher resolutions. It was used for the topographic correction of multi-spectral image in mountainous areas, which had a direct impact on the model-predicted forest canopy height. These problems indicated the importance of the DTM.

In the validation of the predicted forest canopy height, only 20 survey plots were used. This was determined by the rough forest type classification and the lack of other validation data, such as LiDAR in the study area. Our next work will attempt to conduct a more detailed forest type classification by using multi-temporal or hyperspectral data. We also plan to apply this method to areas covered by LiDAR data for a more comprehensive assessment. Researchers familiar with LiDAR, especially airborne LiDAR, may not be satisfied with the accuracy of the proposed method. However, we would like to clarify that our goal was to develop a complementary approach to existing techniques, and the complex terrain in the study area was also a big challenge. The method is expected to achieve better accuracy in areas where the composition of forests is simpler and the terrain is gentler. The results in this study could be considered as a conservative assessment of the proposed method.

The use of the ZY-3 and SRTMGL1 data could facilitate the forest canopy height mapping over large regions. The processing steps mentioned in this study were important for solving the problems caused by the resolution differences between data, and for suppressing the interferences from rugged terrain. In this study, the slope difference layer was very important. Multi-resolution segmentation also played a significant role in refining the crude CHM. If we had not used the average slope of segments as a criterion, the R^2 of the out-of-bag estimation would not have exceeded 0.36, and the RMSE would have been larger than 4 m.

6. Conclusions

This study explored the feasibility of mapping forest canopy height in a mountainous area of Hubei Province, Central China using the China's stereo mapping satellite ZiYuan-3 (ZY-3) data, the Shuttle Radar Topography Mission global 1 arc second data (SRTMGL1), and the Landsat 8 Operational Land Imager (OLI) surface reflectance data. We found that the crude CHM, derived from the ZY-3 and SRTMGL1, contained information on the heights of surface objects, but was contaminated by many erroneous values. The slope difference and the average slope of segments criteria played important roles in refining the crude CHM. The near infrared band and green band of the Landsat 8 data were the most important variables in the Random Forest model. The evaluation of the model using out-of-bag estimation showed an R^2 of 0.53 and a RMSE of 3.28 m. The predicted forest canopy height responded to forest types, altitude, and human activity. We validated the predicted forest canopy height using the mean forest height measured in the field survey plots. The validation showed an R^2 of 0.62 and a RMSE of 2.64 m. It will be easy to combine the proposed method with other techniques. For example, a DSM derived from the X-band SAR images through radargrammetry can be used as a substitute for the ZY-3 DSM in this study. Combining the ZY-3 DSM, with a ALS-derived DTM may improve the quality of the CHM, simplify the refining process and increase the number of training samples. Forest heights derived from the spaceborne LiDAR data, that do not provide full coverage, such as those from the ICESat-2 and GEDI missions, can be used as a substitute for the refined CHM samples in this study to establish and extrapolate the regression model in combination with the Landsat 8 data.

Author Contributions: Data curation, M.L., Y.D., and X.N.; Funding acquisition, C.C.; Investigation, M.L.; Methodology, M.L.; Project administration, C.C., Y.D., and X.N.; Resources, C.C. and Y.D.; Software, Y.D.; Supervision, C.C.; Writing-original draft, M.L.; Writing-review & editing, C.C., and X.N.

Funding: This research was funded by the National Key Research and Development Program of China grant number 2016YFB0501505.

Acknowledgments: Landsat surface reflectance product courtesy of the U.S. Geological Survey. SRTMGL1 Version 3 product was retrieved from the online Data Pool, courtesy of the NASA Land Processes Distributed Active Archive Center (LP DAAC), USGS/Earth Resources Observation and Science (EROS) Center.

Conflicts of Interest: The authors declare no conflict of interest.

Appendix A

We sorted all of the samples, based on their refined CHM values and segmented them at 0.5 m intervals. We then averaged the refined CHM and predictor variables in each interval (i.e., piecewise averaging) and created scatter plots to demonstrate the relationship between them. For the forest type, we counted the number of samples instead of averaging. Figure A1 shows the scatter plots of the piecewise averaged predictor variables and refined CHM. Data points with refined CHM less than 3 m or larger than 23 m were discrete, because these intervals contained few samples.

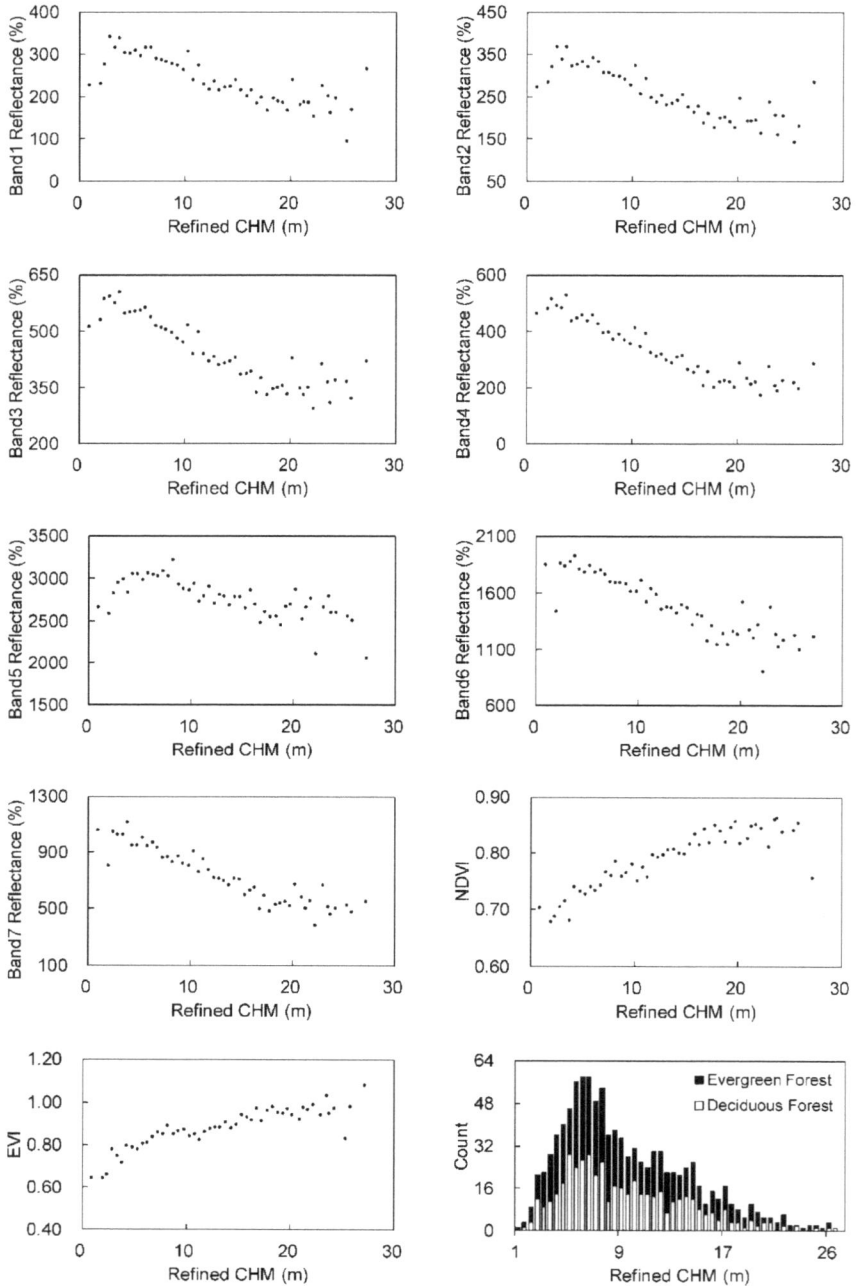

Figure A1. Scatter plots of the piecewise averaged predictor variables and refined CHM.

References

1. Lefsky, M.A.; Keller, M.; Pang, Y.; de Camargo, P.B.; Hunter, M.O. Revised method for forest canopy height estimation from Geoscience Laser Altimeter System waveforms. *J. Appl. Remote Sens.* **2007**, *1*, 1–18.
2. Lefsky, M.A. A global forest canopy height map from the Moderate Resolution Imaging Spectroradiometer and the Geoscience Laser Altimeter System. *Geophys. Res. Lett.* **2010**, *37*, 1–5. [CrossRef]
3. Simard, M.; Pinto, N.; Fisher, J.B.; Baccini, A. Mapping forest canopy height globally with spaceborne lidar. *J. Geophys. Res. Biogeosci.* **2011**, *116*, 1–12. [CrossRef]
4. Roussel, J.R.; Caspersen, J.; Béland, M.; Thomas, S.; Achim, A. Removing bias from LiDAR-based estimates of canopy height: Accounting for the effects of pulse density and footprint size. *Remote Sens. Environ.* **2017**, *198*, 1–16. [CrossRef]
5. Vastaranta, M.; Holopainen, M.; Karjalainen, M.; Kankare, V.; Hyyppa, J.; Kaasalainen, S.; Hyyppa, H. SAR radargrammetry and scanning LiDAR in predicting forest canopy height. *IEEE Int. Geosci. Remote Sens. Symp.* **2012**, *53*, 6515–6518.
6. Balzter, H.; Rowland, C.S.; Saich, P. Forest canopy height and carbon estimation at Monks Wood National Nature Reserve, UK, using dual-wavelength SAR interferometry. *Remote Sens. Environ.* **2007**, *108*, 224–239. [CrossRef]
7. Ghasemi, N.; Tolpekin, V.; Stein, A. A modified model for estimating tree height from PolInSAR with compensation for temporal decorrelation. *Int. J. Appl. Earth Obs.* **2018**, *73*, 313–322. [CrossRef]
8. Balenovic, I.; Seletkovic, A.; Pernar, R.; Jazbec, A. Estimation of the mean tree height of forest stands by photogrammetric measurement using digital aerial images of high spatial resolution. *Ann. For. Res.* **2015**, *58*, 125–143. [CrossRef]
9. Herrero, H.M.; Felipe, G.B.; Belmar, L.S.; Hernandez, L.D.; Rodriguez, G.P.; Gonzalez, A.D. Dense Canopy Height Model from a low-cost photogrammetric platform and LiDAR data. *Trees-Struct. Funct.* **2016**, *30*, 1287–1301. [CrossRef]
10. Zhou, T.; Popescu, S.C.; Krause, K.; Sheridan, R.D.; Putman, E. Gold—A novel deconvolution algorithm with optimization for waveform LiDAR processing. *ISPRS-J. Photogramm. Remote Sens.* **2017**, *129*, 131–150. [CrossRef]
11. Anderson, J.; Martin, M.E.; Smith, M.L.; Dubayah, R.O.; Hofton, M.A.; Hyde, P.; Peterson, B.E.; Blair, J.B.; Knox, R.G. The use of waveform lidar to measure northern temperate mixed conifer and deciduous forest structure in New Hampshire. *Remote Sens. Environ.* **2006**, *105*, 248–261. [CrossRef]
12. Boudreau, J.; Nelson, R.F.; Margolis, H.A.; Beaudoin, A.; Guindon, L.; Kimes, D.S. Regional aboveground forest biomass using airborne and spaceborne LiDAR in Québec. *Remote Sens. Environ.* **2008**, *112*, 3876–3890. [CrossRef]
13. Huang, H.; Liu, C.; Wang, X.; Biging, G.S.; Chen, Y.; Yang, J.; Gong, P. Mapping vegetation heights in China using slope correction ICESat data, SRTM, MODIS-derived and climate data. *ISPRS-J. Photogramm. Remote Sens.* **2017**, *129*, 189–199. [CrossRef]
14. Bolton, D.K.; Coops, N.C.; Wulder, M.A. Investigating the agreement between global canopy height maps and airborne LiDAR derived height estimates over Canada. *Can. J. Remote Sens.* **2013**, *39*, S139–S151. [CrossRef]
15. Hilker, T.; Wulder, M.A.; Coops, N.C. Update of forest inventory data with LiDAR and high spatial resolution satellite imagery. *Can. J. Remote Sens.* **2008**, *34*, 5–12. [CrossRef]
16. Hansen, M.C.; Potapov, P.V.; Goetz, S.J.; Turubanova, S.; Tyukavina, A.; Krylov, A.; Kommareddy, A.; Egorov, A. Mapping tree height distributions in Sub-Saharan Africa using Landsat 7 and 8 data. *Remote Sens. Environ.* **2016**, *185*, 221–232. [CrossRef]
17. Garestier, F.; Dubois-Fernandez, P.C.; Guyon, D.; Le Toan, T. Forest biophysical parameter estimation using L- and P-band polarimetric SAR data. *IEEE Trans. Geosci. Remote Sens.* **2009**, *47*, 3379–3388. [CrossRef]
18. Karjalainen, M.; Kankare, V.; Vastaranta, M.; Holopainen, M.; Hyyppä, J. Prediction of plot-level forest variables using TerraSAR-X stereo SAR data. *Remote Sens. Environ.* **2012**, *117*, 338–347. [CrossRef]
19. Capaldo, P.; Nascetti, A.; Porfiri, M.; Pieralice, F.; Fratarcangeli, F.; Crespi, M.; Toutin, T. Evaluation and comparison of different radargrammetric approaches for Digital Surface Models generation from COSMO-SkyMed, TerraSAR-X, RADARSAT-2 imagery: Analysis of Beauport (Canada) test site. *ISPRS-J. Photogramm. Remote Sens.* **2015**, *100*, 60–70. [CrossRef]

20. Solberg, S.; Astrup, R.; Breidenbach, J.; Nilsen, B.; Weydahl, D. Monitoring spruce volume and biomass with InSAR data from TanDEM-X. *Remote Sens. Environ.* **2013**, *139*, 60–67. [CrossRef]

21. Zhang, Y.; He, C.; Xu, X.; Chen, D. Forest vertical parameter estimation using PolInSAR imagery based on radiometric correction. *ISPRS Int. J. Geo-Inf.* **2016**, *5*, 186. [CrossRef]

22. Neuenschwander, A.; Pitts, K. The ATL08 land and vegetation product for the ICESat-2 Mission. *Remote Sens. Environ.* **2019**, *221*, 247–259. [CrossRef]

23. Qi, W.; Dubayah, R.O. Combining Tandem-X InSAR and simulated GEDI lidar observations for forest structure mapping. *Remote Sens. Environ.* **2016**, *187*, 253–266. [CrossRef]

24. Le Toan, T.; Quegan, S.; Davidson, M.W.J.; Balzter, H.; Paillou, P.; Papathanassiou, K.; Plummer, S.; Rocca, F.; Saatchi, S.; Shugart, H.; et al. The BIOMASS mission: Mapping global forest biomass to better understand the terrestrial carbon cycle. *Remote Sens. Environ.* **2011**, *115*, 2850–2860. [CrossRef]

25. Moreira, A.; Krieger, G.; Hajnsek, I.; Papathanassiou, K.; Younis, M.; Lopez-Dekker, P.; Huber, S.; Villano, M.; Pardini, M.; Eineder, M.; et al. Tandem-L: A Highly Innovative Bistatic SAR Mission for Global Observation of Dynamic Processes on the Earth's Surface. *IEEE Geosci. Remote Sens. Mag.* **2015**, *3*, 8–23. [CrossRef]

26. White, J.C.; Wulder, M.A.; Vastaranta, M.; Coops, N.C.; Pitt, D.; Woods, M. The utility of image-based point clouds for forest inventory: A comparison with airborne laser scanning. *Forests* **2013**, *4*, 518–536. [CrossRef]

27. Aguilar, M.A.; Del Mar Saldaña, M.; Aguilar, F.J. Generation and quality assessment of stereo-extracted DSM from GeoEye-1 and WorldView-2 imagery. *IEEE Trans. Geosci. Remote Sens.* **2014**, *52*, 1259–1271. [CrossRef]

28. Immitzer, M.; Stepper, C.; Böck, S.; Straub, C.; Atzberger, C. Use of WorldView-2 stereo imagery and National Forest Inventory data for wall-to-wall mapping of growing stock. *For. Ecol. Manag.* **2016**, *359*, 232–246. [CrossRef]

29. Jayathunga, S.; Owari, T.; Tsuyuki, S. Evaluating the Performance of Photogrammetric Products Using Fixed-Wing UAV Imagery over a Mixed Conifer-Broadleaf Forest: Comparison with Airborne Laser Scanning. *Remote Sens.* **2018**, *10*, 187. [CrossRef]

30. Véga, C.; St-Onge, B. Height growth reconstruction of a boreal forest canopy over a period of 58 years using a combination of photogrammetric and lidar models. *Remote Sens. Environ.* **2008**, *112*, 1784–1794. [CrossRef]

31. Järnstedt, J.; Pekkarinen, A.; Tuominen, S.; Ginzler, C.; Holopainen, M.; Viitala, R. Forest variable estimation using a high-resolution digital surface model. *ISPRS-J. Photogramm. Remote Sens.* **2012**, *74*, 78–84. [CrossRef]

32. Pearse, G.D.; Dash, J.P.; Persson, H.J.; Watt, M.S. Comparison of high-density LiDAR and satellite photogrammetry for forest inventory. *ISPRS-J. Photogramm. Remote Sens.* **2018**, *142*, 257–267. [CrossRef]

33. Gan, J.; Ge, J.; Liu, Y.; Wang, Z.; Wu, X.; Ding, S. Forest ecosystem quality change over ten years (2000–2010) in the Three Gorges Dalaoling Nature Reserve. *Plant Sci. J.* **2015**, *33*, 766–774.

34. Wang, T.; Zhang, G.; Li, D.; Tang, X.; Jiang, Y.; Pan, H.; Zhu, X.; Fang, C. Geometric accuracy validation for ZY-3 satellite imagery. *IEEE Geosci. Remote Sens. Lett.* **2014**, *11*, 1168–1171. [CrossRef]

35. Li, D. China's first civilian three-line-array stereo mapping satellite ZY-3. *Acta Geodaetica et Cartographica Sinica* **2012**, *41*, 317–322.

36. Pan, H.; Zhang, G.; Tang, X.; Wang, X.; Zhou, P.; Xu, M.; Li, D. Accuracy analysis and verification of ZY-3 products. *Acta Geodaetica et Cartographica Sinica* **2013**, *42*, 738–744.

37. Ni, W.; Sun, G.; Ranson, K.J.; Pang, Y.; Zhang, Z.; Yao, W. Extraction of ground surface elevation from ZY-3 winter stereo imagery over deciduous forested areas. *Remote Sens. Environ.* **2015**, *159*, 194–202. [CrossRef]

38. Farr, T.G.; Kobrick, M. Shuttle radar topography mission produces a wealth of data. *Eos* **2000**, *81*, 583–585. [CrossRef]

39. Farr, T.G.; Rosen, P.A.; Caro, E.; Crippen, R.; Duren, R.; Hensley, S.; Kobrick, M.; Paller, M.; Rodriguez, E.; Roth, L.; et al. The Shuttle Radar Topography Mission. *Rev. Geophys.* **2007**, *45*, 1–43. [CrossRef]

40. Kobrick, M. On the toes of giants-How SRTM was born. *Photogramm. Eng. Remote Sens.* **2006**, *72*, 206–210.

41. Nikolakopoulos, K.G.; Kamaratakis, E.K.; Chrysoulakis, N. SRTM vs ASTER elevation products. Comparison for two regions in Crete, Greece. *Int. J. Remote Sens.* **2006**, *27*, 4819–4838. [CrossRef]

42. Rosen, P.A.; Hensley, S.; Joughin, I.R.; Li, F.K.; Madsen, S.N.; Rodríguez, E.; Goldstein, R.M. Synthetic aperture radar interferometry. *Proc. IEEE* **2000**, *88*, 333–380. [CrossRef]

43. Hu, Z.; Peng, J.; Hou, Y.; Shan, J. Evaluation of recently released open global digital elevation models of Hubei, China. *Remote Sens.* **2017**, *9*, 262. [CrossRef]

Forests **2019**, *10*, 105

44. Krieger, G.; Moreira, A.; Fiedler, H.; Hajnsek, I.; Werner, M.; Younis, M.; Zink, M. TanDEM-X: A satellite formation for high-resolution SAR interferometry. *IEEE Trans. Geosci. Remote Sens.* **2007**, *45*, 3317–3341. [CrossRef]

45. De Oliveira, C.G.; Paradella, W.R.; da Silva, A.D.Q. Assessment of radargrammetric DSMs from TerraSAR-X Stripmap images in a mountainous relief area of the Amazon region. *ISPRS-J. Photogramm. Remote Sens.* **2011**, *66*, 67–72. [CrossRef]

46. Su, Y.; Guo, Q. A practical method for SRTM DEM correction over vegetated mountain areas. *ISPRS-J. Photogramm. Remote Sens.* **2014**, *87*, 216–228. [CrossRef]

47. Vermote, E.; Justice, C.; Claverie, M.; Franch, B. Preliminary analysis of the performance of the Landsat 8/OLI land surface reflectance product. *Remote Sens. Environ.* **2016**, *185*, 46–56. [CrossRef]

48. Zhu, Z.; Woodcock, C.E. Object-based cloud and cloud shadow detection in Landsat imagery. *Remote Sens. Environ.* **2012**, *118*, 83–94. [CrossRef]

49. Wang, Y.; Li, G.; Ding, J.; Guo, Z.; Tang, S.; Wang, C.; Huang, Q.; Liu, R.; Chen, J.M. A combined GLAS and MODIS estimation of the global distribution of mean forest canopy height. *Remote Sens. Environ.* **2016**, *174*, 24–43. [CrossRef]

50. Teillet, P.M.; Guindon, B.; Goodenough, D.G. On the slope-aspect correction of multispectral scanner data. *Can. J. Remote Sens.* **1982**, *8*, 84–106. [CrossRef]

51. Breiman, L. Random forest. *Mach. Learn.* **2001**, *45*, 5–32. [CrossRef]

52. Congalton, R.G. A review of assessing the accuracy of classifications of remotely sensed data. *Remote Sens. Environ.* **1991**, *37*, 35–46. [CrossRef]

forests

MDPI

Article

Application of Terrestrial Laser Scanner to Evaluate the Influence of Root Collar Geometry on Stump Height after Mechanized Forest Operations

Eric R. Labelle [1,*], **Joachim B. Heppelmann** [1] and **Herbert Borchert** [2]

1   Department of Ecology and Ecosystem Management, Technical University of Munich,
    Hans-Carl-von-Carlowitz-Platz 2, D-85354 Freising, Germany; joachim.heppelmann@hswt.de
2   Bayerische Landesanstalt für Wald und Forstwirtschaft, Hans-Carl-von-Carlowitz-Platz 1,
    D-85354 Freising, Germany; herbert.borchert@lwf.bayern.de
*   Correspondence: eric.labelle@tum.de; Tel.: +49-816-171-4760

Received: 26 October 2018; Accepted: 14 November 2018; Published: 15 November 2018

check for
updates

Abstract: The height of tree stumps following mechanized forest operations can be influenced by machine-, tree-, terrain-, and operator-related characteristics. High stumps may pose different economic and technical disadvantages. Aside from a reduction in product recovery (often associated with sawlog potential), leaving high stumps can complicate future entries if smaller equipment with low ground clearance is used, particularly in the case where new machine operating trails are required. The objective of this exploratory study was to examine if correlations existed between the height of tree stumps following mechanized harvesting and the shape of the above-ground root collar, stump diameter, and distance to the machine operating trail. In total, 202 sample stumps of Norway spruce (*Picea abies* (L.) Karst.) and the surrounding terrain were scanned with a terrestrial laser scanner. The collected data was processed into a 3D-model and then analyzed. Stump height was compared with different characteristics such as stump diameter at the cut surface, distance to the machine operating trail, number of visible root flares per stump, and the root collar. The number of root flares per stump had a positive influence on stump diameter and height, showing a general trend of increasing diameter and height with the increasing number of root flares. Root angles also had an influence on the stump diameter. The diameter of a stump and the shape of the root collar at the cut surface together had a significant effect on stump height and the model reported explained half of the variation of stump heights. Taken together, these findings suggest that other factors than the ones studied can also contribute in influencing stump height during mechanized harvesting operations. Further investigations, including pre- and post-harvest scans of trees selected for removal, are warranted.

Keywords: stump diameter; stump height; harvester; product recovery; Norway spruce

1. Introduction

A recent change in German forestry has been the increased use of mechanized forest operations, particularly when considering the cut-to-length (CTL) method [1]. In a fully mechanized CTL system, a harvester is commonly used to fell, delimb, and buck the stem into logs of various lengths directly on the machine operating trail after which a forwarder is used to transport the processed logs to the roadside or to a landing area. Harvesters were first introduced in Germany during the 1990 large-scale wind-throws, which uprooted about 72.5 million m^3 of wood and required a sudden widespread increase of mechanized operations to safely and efficiently harvest the unusually large amount of wood [2]. Nowadays, hundreds of harvesters are in use in Germany and they are the most common machines encountered in mechanized operations [3]. The proven benefits of fully-mechanized harvesting systems are the increased work safety, the combination of different working steps into one process, and increased monetary profits via higher productivity and efficiency [4]. Currently, about 60% of all wood harvested from German public forests is with mechanized forest operations and from this proportion, the entirety is with the cut-to-length (CTL) method [1]. While tree size and shape have less influence on the feasibility of chainsaw operations, harvesters can reach their operational limit with large diameter trees or with trees exhibiting relatively large diameters and complex root systems, thus potentially leaving high stumps in the forest. Such stumps can be unwanted obstacles making it difficult to maneuver machines through the forest or transport wood without damaging residual trees. Additionally, the Food and Agriculture Organization of the United Nations suggests stump height should be as low as practicable and identifies 30 cm or lower as a preferable threshold to maximize merchantable volume [5]. With advancements in machine design and with specific requirements from third-party certification programs, it is not uncommon to see alterations in the design/layout of machine operating trails. Any modification in spatial alignment or spacing of machine operating trails will entail that some new trails have to be created in areas that were previously in the leave strip (area adjacent to two trails). It should also be of interest to harvest as much woody biomass as possible and produce logs of longer length, especially in the lower section of the tree, where the diameter is the greatest and normally presents fewer irregularities and, thus, providing higher returns.

Concerning woody biomass, measurements done throughout the last century were mostly performed manually with simple techniques, tables, and often relying on estimations rather than actual measurements. The terrestrial laser scanner (TLS), on the other hand, provides a relatively easy, fast, and precise method to scan and measure forest structures by sending out an infrared laser beam scanning the surrounding environment on the millimeter scale. This is of particular importance when the scanning subjects have an irregular shape, such as tree stumps. The TLS is a non-intrusive and non-destructive measurement device that allows for repeated measurements at the exact same locations. Initially developed to address issues in the industrial building sector, the use of TLS in forestry settings is increasing in frequency. TLS data have already been used to estimate forest structures such as sub-canopy architecture [6], leaf area index [7–9], tree height, diameter and diameter at breast height [10–15], and specific tree properties, such as stem volume [14] and crown characteristics [16–19]. Regarding stump measurements, TLS was used to assess above-ground stump biomass and the associated indirect emission of bioenergy [20]. The most noticeable advantages of this system are the possibilities of capturing data without disturbing the forest environment. Once recorded, the data can be used to measure variables of interest without the need to revisit the study area if further information is missing. Despite the research and advancements listed above, very limited research has been completed with TLS to measure stump characteristics in order to investigate the influence of above-ground root collar on stump height following mechanized felling.

To allow mechanized forest operations to be as effective as possible, it is necessary to identify and understand physical barriers that might influence the height at which a tree can be cut and felled. Such a barrier might be the stump diameter. Each harvesting head has a limited range of stem diameters which can be cut. The diameter can be limited either by the opening of the feed-rollers or the length of the saw bar. If the diameter close to the ground is too large, the harvesting head can be lifted upwards until a suitable diameter is reached, thereby creating a higher stump. Striking root flares might also hinder an adjustment of the harvesting head close to the ground. Together with an increasing distance from the machine, the lifting force of the harvester boom decreases. Some harvesters can hardly lift the harvesting head alone if the boom is fully extended. In this case, a proper arrangement of the harvesting head at the tree is difficult. At greater distances, the visibility of operators can also be hindered. Both might contribute to higher stumps with an increasing distance between the machine operating trail and the tree.

In the scope of this study, we intended to test the following hypotheses:

Hypothesis 1. *We anticipate a correlation between stump height and stump diameter, which will lead to higher stumps as the diameter at the cut surface is increasing.*

Hypothesis 2. *We anticipate a correlation between the distance from a stump to the machine operating trail and stump height resulting in higher stumps as the distance between the stump and trail is increasing.*

Hypothesis 3. *We also anticipate a correlation between stump diameter and root collar, which will lead to a higher stump diameter as the root angles are getting flatter and are, therefore, also causing higher stumps.*

Through the use of a TLS, the objective of this study was to find possible correlations between stump height and physical parameters, such as stump diameter, above-ground root collar, and distance to the machine operating trail following mechanized operations. In this context, we defined a root collar as the above-ground widening of the stump beyond its natural taper.

Other factors may also affect stump height. This can be the power of the harvester, the skill of the operator or the tree species being harvested. These factors can only be analyzed when a large number of different logging operations are observed. However, we investigated the stumps of only one logging operation and could, therefore, not address the above-mentioned factors.

2. Materials and Methods

2.1. Research Site and Experimental Design

Research sites were located in southern Germany near the town of Freising (Figure 1A). In total, 103 stumps were measured in the forest district Rappenberg (48°24′25.2″ N and 11°42′19.6″ E) and 99 stumps in the neighboring district Wippenhausereinfang (48°25′33.1″ N and 11°41′13.6″ E). The sites were chosen because of the fully-mechanized operations and their proximity to the Technical University of Munich. Mechanized harvesting was performed in February and March 2015 by an experienced entrepreneur commissioned by the Bavarian State Forest Enterprise to harvest trees following a wind-throw. This calamity triggered an opportunity to perform additional thinning operations near the affected areas. However, only the trees that showed no sign of wind-throw were analyzed in this study. To exclude the influences of different mechanical specifications, the study area only involved the sections harvested by the same machine and operator.

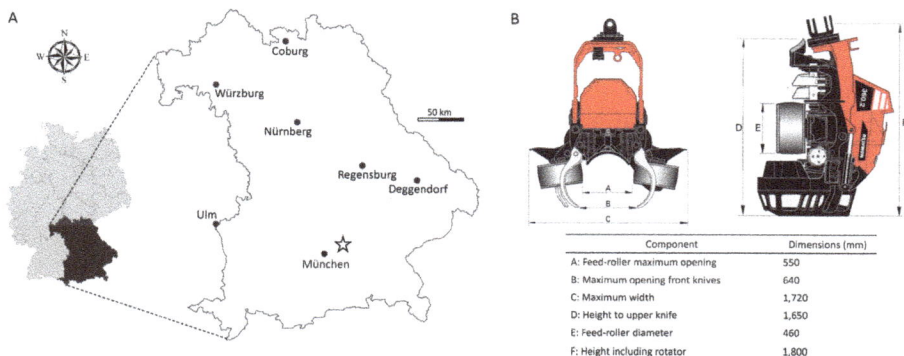

Component	Dimensions (mm)
A: Feed-roller maximum opening	550
B: Maximum opening front knives	640
C: Maximum width	1,720
D: Height to upper knife	1,650
E: Feed-roller diameter	460
F: Height including rotator	1,800

Figure 1. (**A**). Location of research sites depicted by star symbol within the State of Bavaria, Germany; (**B**). Top and side-view of a Komatsu 360.2 harvesting head along with dimensions of the main components [21].

The forest is mainly stocked with Norway spruce (*Picea abies*) (80%) and European beech (*Fagus sylvatica* L.) (15%) mixed with a few Douglas fir (*Pseudotsuga menziesii* (Mirb.) Franco), larch (*Larix decidua* Mill.) and oak (*Quercus* L.) trees constituting the remaining 5%. The mechanized portion of the felling and our associated measurements were performed in a Norway spruce-dominated stand. For this study, stumps originating from 202 harvester-felled Norway spruce trees with varying diameters along with the corresponding machine operating trails and surrounding forest areas were scanned with a TLS. The only stumps targeted were those that were (i) not affected by the wind-throw and, thus, possessed an undisturbed root collar and (ii) stumps originating from the current harvest. Due to the very limited spatial extent of the test site, similar stand and terrain conditions were present.

All of the 202 trees corresponding to our sampled stumps had previously been harvested and processed with a Valmet 921 (Komatsu Forest AB, Umea, Sweden) six-wheeled single-grip harvester with a 10 m long boom equipped with a Komatsu forest 360.2 harvesting head (Komatsu Forest AB, Umea, Sweden) (Figure 1B). This particular harvesting head had a maximum opening of the upper delimbing knives of 64 cm and a maximum feed roller opening of 55 cm [21].

2.2. Instrumentation and Sampling

Initial field tests relied on conventional measurements of stump characteristics using rulers, string and a builder's level. However, severe difficulties were encountered when measuring the angles of the root flares since it was almost impossible to have a fixed reference point perpendicular to the ground and this for each 5 cm layer. For this reason, a TLS was instead chosen to collect field data consisting of 3D point clouds. The settings (resolution and quality) selected for this experiment corresponded to a point distance of 7.8 mm in a scan distance of 10 m with an average scan time of 190 seconds (not including site preparation and TLS setup) [22]. With these recording parameters, high accuracy was achieved while keeping the amount of data and scanning time appropriate.

Preliminary test scans indicated that moss and branches on the surface and sides of stumps could cause difficulties during data processing. Therefore, prior to the start of a scan, all sample stumps were cleared of moss and loose material (branches, leaves, dead material, etc.) from the cut surface down to the soil layer with the use of steel brushes, a procedure that took on average one minute per stump. Surrounding under-growth vegetation was also removed if it was deemed to hinder the upcoming scanning campaign. To facilitate the identification of the measured stumps within the point cloud, a wooden stake was inserted near each sampled stump prior to scanning.

Spheres of 145 mm diameter, automatically detectable by the processing software, were used in the field to facilitate merging point clouds originating from different scans. It was necessary that at least three spheres remained in the same location between two adjacent scans, thus permitting the software to triangulate the positions of the scanner. Prior to a scan, all positions of the spheres were verified by line-of-sight to ensure the scanner could detect the spheres from its current position.

To allow for exact measurements, the TLS was leveled at every scan position with the use of an adjustable tripod. Under ideal scanning conditions, a stump was scanned from three sides. One scan was always performed directly from the machine operating trail so that the horizontal distance between the trail centerline and the target stump could be measured. Whenever possible, the scans followed a diagonal pattern on one or both sides of the machine operating trail (Figure 2). In areas where dense vegetation or complex root systems were present, scanning frequency was increased accordingly to assure that all required data was captured.

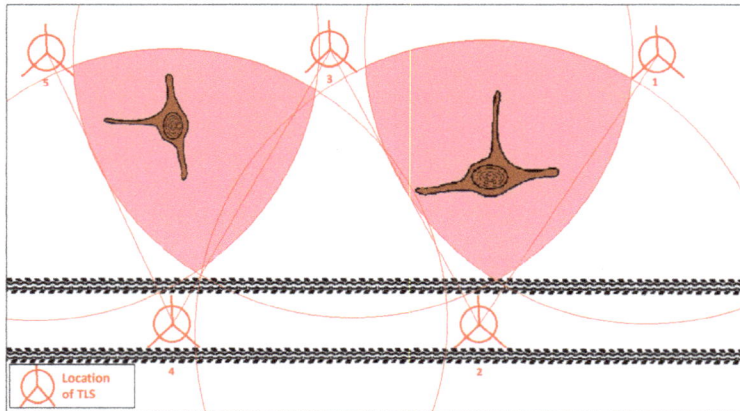

Figure 2. Schematic of an optimal scanning pattern to allow visualization and measurements of the stumps and machine operating trail.

2.3. Data and Statistical Analyses

Following the field campaign, data was examined using a point cloud processing, visualizing, and analyzing software. To perform exact measurements of the stump architecture, it was first necessary to create a mesh on the surface of the stump rather than try to obtain required measurements via the point cloud. This procedure, performed for all target stumps and corresponding machine operating trails, consisted of creating a triangular irregular network, which converted the point clouds into surfaces.

This study focused on characterizing four main parameters; stump diameter, stump height, root collar geometry, and distance to machine operating trail. Since it was of interest to understand the change in stump diameter from the cut surface down towards the soil surface, three different types of stump diameters will be discussed (diameter at cut surface, diameter at the 5 cm segment, and diameter at the 10 cm segment). All stump diameters have their starting point or starting horizontal plane on the top surface (cut surface) of the stump, which was in contact with the saw of the harvesting head. Unless otherwise specified, the diameter at cut surface is the origin (0 cm) and from this position, measurements were recorded in a vertical downward direction towards the ground in 5 cm increments. As an example, the further described 5 cm segment was the segment from the edge of the cut surface to a plane 5 cm below the cut surface (Figure 3E). Methods used to quantify stump characteristics are summarized below.

- Diameters: Similar to the method used by [23,24], average stump diameters (mm accuracy), irrespective of their vertical position on the stump, were calculated using two outside bark diameter measurements (shortest and longest) extending through the geometric center of the stump (Figure 3C).
- Heights: Stump heights were measured with mm accuracy from the ground level on the high side (in reference to the ground) of the stump to the height at the geometric center at the cut surface [23] (Figure 3D).
- Root collar: To appropriately describe the root collar, every angle between the created vertical segments on the stump was measured to a horizontal line. The first angle was always the angle from the edge of the stump cut surface down to the 5 cm line. The angles were measured on top of the roots in a downward direction towards the ground (Figure 3F). Root collar was assessed in three segments; cut surface (0 cm) to 5 cm, 5 to 10 cm, and 10 to 15 cm. For example, a 90° angle would correspond to a root not extending beyond the natural taper of the stem.

- Number of root flares per stump: Above-ground root flares were defined as visually discernable parts of a stump beyond its natural taper, which continuously extended down to the soil layer (as shown in Figure 3C).
- Distance to machine operating trail: The distance from a stump to the machine operating trail was measured from the geometric center of the stump to the centerline of the operation trail in a 90° angle. To allow for this type of measurement, a line was generated following the middle of the trail and used as a reference for the 90° measurements (Figure 3B).

The measurements described above only consider the shape of the top 15 cm of a stump as expressed from the cut surface downwards. In order to consider the shape of the entire stump, the diameters and root angles of a subsample of 100 stumps (randomly selected from the entire 202 stump data set) were also measured with the point of origin now beginning from the ground surface upwards in 5 cm layers until reaching the cut surface. In this subset, minimum diameters were measured similar to the functioning of a tree caliper. Thus, smaller diameters only caused by grooves or protrusions on a stump did not influence the caliper-type diameter measurements. The complimentary analysis was performed to better link the stump characteristics to the functioning of a harvesting head as it is being positioned at the base of a tree.

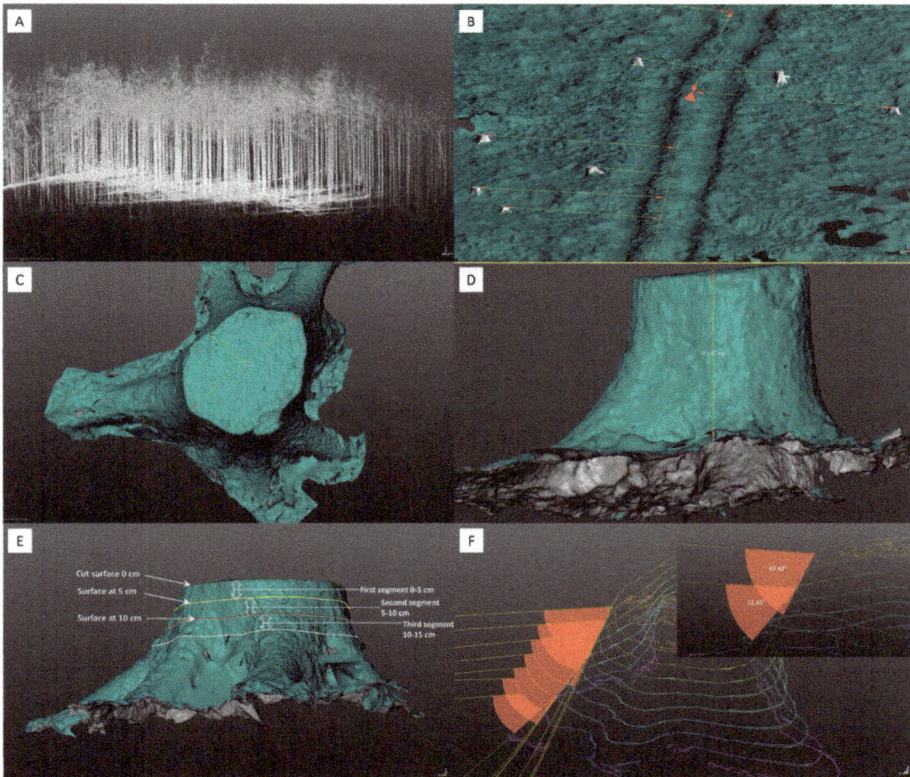

Figure 3. (A) Point cloud of the stand; (B) isometric-view of machine operating trail and stumps; (C) top-view of stump depicting the location of diameter measurements; (D) side-view of stump depicting height measurement; (E) side-view of stump showing the first three 5 cm segments and corresponding surfaces, and; (F) isometric-view of the angles measured for each layer and each root flare.

For all ensuing measurements, data was transferred and analyzed in Minitab 17 and R Statistics. All response variables (diameters at different heights, height, distance to machine operating trail) were first tested for normality with the Anderson-Darling normality test. One-way ANOVAs were performed in Minitab with the diameter and height set as response variables and the number of roots per tree as the factor. R Statistics was used to obtain Pearson correlations. Additionally, a multiple linear regression analysis (ordinary least squares method) was performed in SAS 9.3 where the stump height was the dependent variable and the minimum diameter and average root angles at the cut surface and the number of root flares and the distance to the machine operating trail used as predictor variables. Outliers were identified by Cook's D, which combines information on the residual and leverage [25]. The normality of the residuals was proved by the Shapiro-Wilk W test for normality and homoscedasticity was tested by performing the White test. Lastly, multicollinearity was tested by looking at the variance inflation factor.

Another linear regression was performed with the data from the subsample of 100 stumps where measurements originated from the ground. The minimum diameter and the average of the angles of all roots were the explanatory variables and the distance to the cut surface the dependent variable. Here the average of the root angles of successive layers was assigned to the diameter between both segments. In addition, a quantile regression for this model was executed since they are useful in applications where extremes are important [26]. Therefore, the model should reveal if the variation of stump heights at the lower end are limited by stump diameter and the root collar and ultimately shed some light on if a threshold on the stump height imposed by the stump shape can be determined. In all statistical tests, a significance level of 0.05 was used.

3. Results

3.1. Stump Characteristics

3.1.1. Diameter and Height

All stump diameters, irrespective of their layers of measurement, (e.g., at cut surface, 5 cm or 10 cm below cut surface) followed a normal distribution based on the Anderson-Darling normality test (Figure 4A–C). At the cut surface (Figure 4A), stump diameters ranged from 13.3 cm to 62.3 cm with an average of 37.9 cm. As Figure 4A–C show, stump diameter increased as the measurement plane approached the soil surface with averages of 37.9, 41.4, and 44.0 cm for the cut surface, 5 cm segment, and 10 cm segment, respectively. At 10 cm below the cut surface, the total number of stumps with a diameter greater than 45 cm was 94 instead of 56 when only considering the diameters at the cut surface. That implies that 38 stumps (19% of the total population) were reaching this diameter in only a 10 cm height difference on the vertical. It was apparent that the average diameter, minimum, and maximum was increasing as the measurement plane moved in a downward direction towards the soil surface. In fact, the diameters of stump increased on average by 9.2% between the 0 and 5 cm segments and by 4.6% between the 5 and 10 cm segments.

Unlike the frequency distribution of stump diameters, which all followed a normal distribution, stump heights were not normally distributed (Figure 4D). There was a higher frequency of stumps with heights between 15 and 25 cm as compared to the normal distribution curve. Stump heights varied between 8.8 cm and 57.1 cm with an average of 30.8 cm.

3.1.2. Root Collar

On average, 3.5 root flares per stump were detectable (Figure 5A). Only seven stumps (3.5%) showed no visible above-ground root flares meaning that the stump maintained somewhat of a cylindrical shape from the cut surface down to the soil layer. Due to the absence of visible root flares, those seven stumps were excluded from all upcoming root-related calculations. In general, most of the stumps (119 out of 195) had three or four root flares per stump, whereas only seven stumps had six or more visible above-ground root flares.

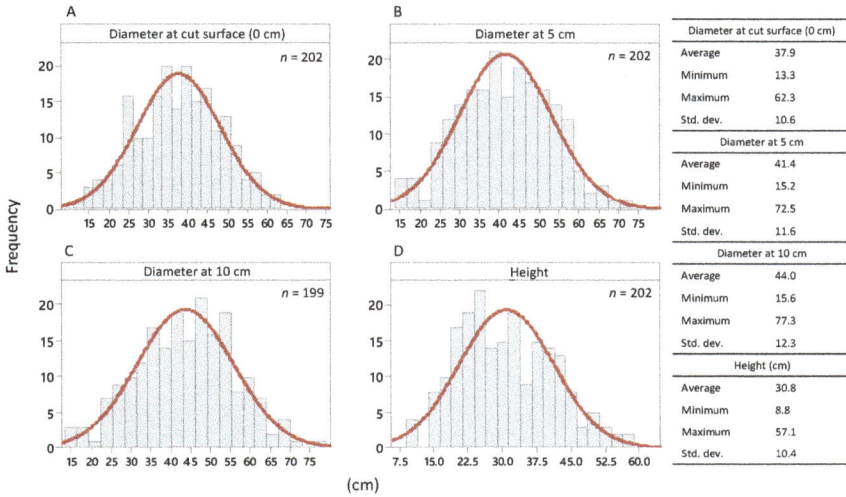

Figure 4. Frequency distributions for (**A**) stump diameters at cut-surface; (**B**) stump diameters at the 5 cm surface; (**C**) stump diameters at the 10 cm surface, and; (**D**) stump heights.

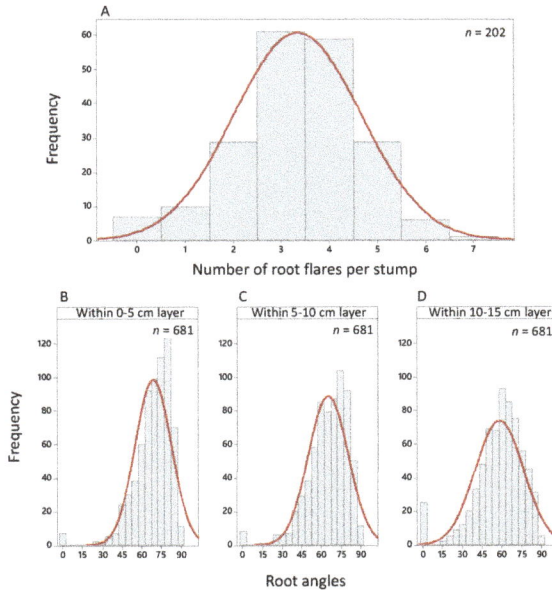

Figure 5. Frequency distributions for (**A**) number of root flares per stump; (**B**) root angles within the 0–5 cm layer; (**C**) root angles within the 5–10 cm layer; and (**D**) root angles within the 10–15 cm layer.

The 202 analyzed stumps provided 675 visually detectable above-ground root flares (as seen from the root collar) (Figure 5B–D). In general, a higher frequency of shallow root angles was measured as the plane used for calculations approached the ground. When focusing on root angles above 72.5°, the frequency of root flares decreased from 46% to 38% and down again to 20% for segments at cut surface, 5–10 cm, and 10–15 cm, respectively.

3.2. Investigation of Relations

3.2.1. Relation between Stump Height/Diameter and Distance to Machine Operating Trail

There did not seem to be any trend between stump diameter and distance to machine operating trail (Figure 6A), a result also supported by a very low Pearson correlation of 0.08 ($p = 0.25$). In fact, 75% of the stumps were located within 8 m from the centerline of the machine operating trails and only 12% were situated beyond 10 m. As the point cloud in Figure 6B shows, there was again no discernable linear correlation between stump height and distance to machine operating trail, as supported by a poor Pearson correlation of 0.06 ($p = 0.40$). There was in fact, high variability in stump height for a respective distance to trail.

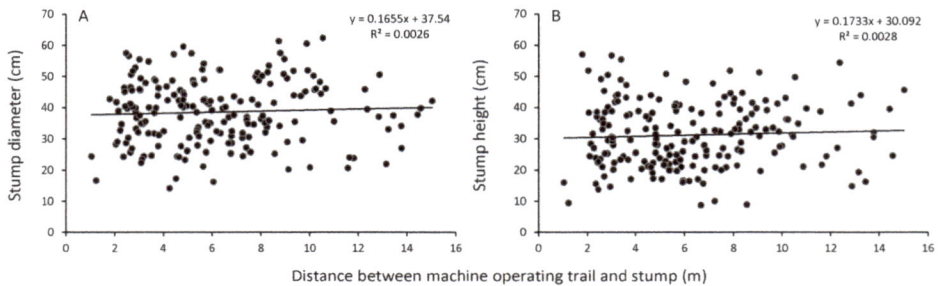

Figure 6. Relation between (**A**) stump diameter and distance to machine operating trail; and (**B**) stump height and distance to machine operating trail.

3.2.2. Relation between Stump Diameter and Stump Height

To determine if stump height was influenced by stump diameter, a possible correlation between both parameters was examined by plotting those two values against each other (Figure 7). To investigate the relationship, a Pearson correlation coefficient of 0.44 ($p = 6.66 \times 10^{-11}$) was calculated indicating that higher stumps tend to have larger diameters at the cut surface. The square of the obtained coefficient indicated that the stump diameter could explain 19% of the variation in stump heights.

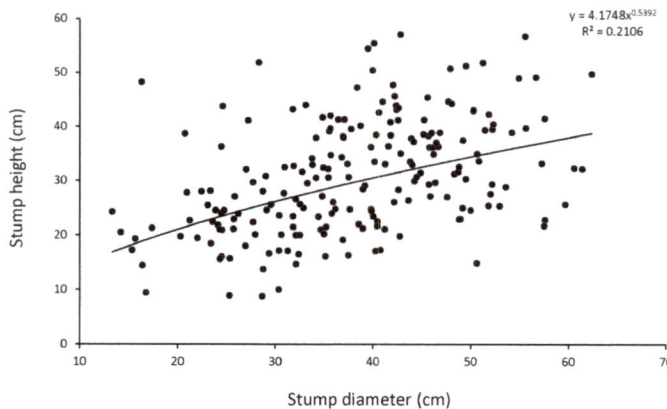

Figure 7. Relation between stump height and stump diameter.

3.2.3. Relation between Number of Root Flares per Stump and Stump Diameter/Height

A statistically significant ($F = 12.40$, $p = 0.000$) positive relationship existed between stump diameter and its corresponding number of root flares (Figure 8A). Average stump diameters increased from 20.4 cm when no above-ground root flares were visible within the root collar to 50.0 cm when six root flares were visible per stump. Stump height was also statistically influenced by the number of root flares per stump ($F = 2.71$, $p = 0.01$) with an increasing stump height as the number of root flares per stump increased (Figure 8B). When performing Fisher pairwise comparisons between the frequency of root flares per stump, statistically significant average stump heights were detected.

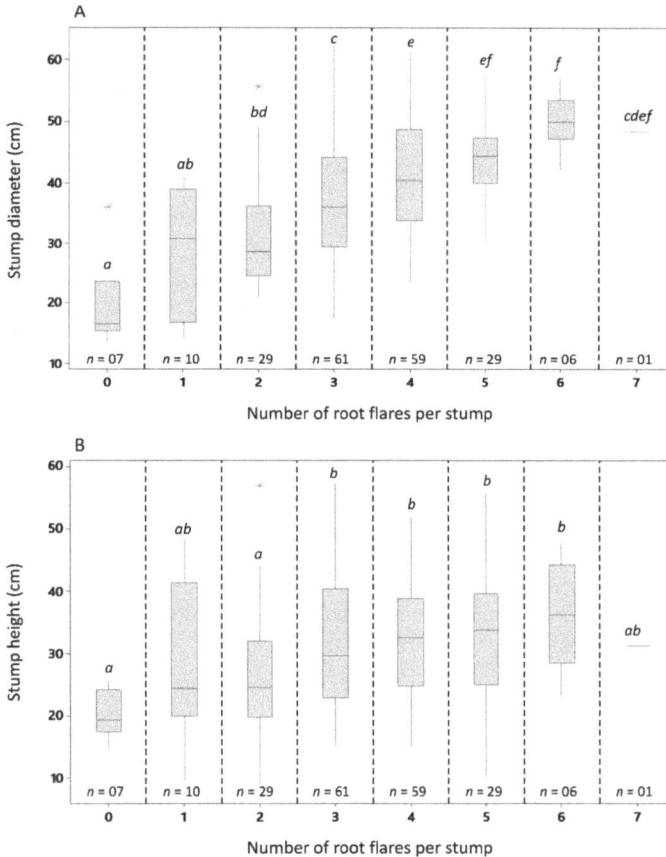

Figure 8. The influence of the number of root flares per stump on (**A**) stump diameter; and (**B**) stump height. Different lower case letters indicate a statistical difference at alpha 0.05 between the number of root flares per stump and stump diameter or height.

3.2.4. Concurrent Effects on Stump Height

The multiple regression model of stump height being predicted by diameter, root angle at the cut surface, number of root flares and distance to the machine operating trail revealed that the last two variable coefficients, were not significantly different from zero. Thus, these variables were excluded from the model, as well as 12 outliers. Both coefficients as well as the intercept of the following model (Equation (1)) were significantly different from zero (Pr > |t| is < 0.0001):

$$Stump\ height\ (mm) = -326.39 + 0.813 \times minimum\ diameter\ (mm) + 4.96 \times root\ angle\ (°) \qquad (1)$$

The positive signs of the predictor variable coefficients were plausible and the adjusted R-square was 0.49. Linearity between the predictor variables and stump height existed and the residuals were normally distributed. A variance inflation factor of 1.2 indicated no collinearity, while the White test indicated homoscedasticity.

The model was also tested with the average diameter and the maximum diameter at the cut surface instead of the minimum diameter. The coefficients were always significantly different from zero. However, the adjusted R-square was slightly lower with 0.46 in case of the average and 0.40 in case of the maximum diameter. In addition, a model was tested with the root angle and the number of root flares being the predictors of the stump height. However, in this instance, the coefficient of the variable root flare number was significantly different from zero but this model gained an adjusted R-square of only 0.14 and, thus, could hardly explain any variation of the stump heights.

3.2.5. Relationship between Diameter and Root Angles and Their Distances to the Cut Surface

The regression analysis of the minimum diameters and average root angles at all 5 cm-segments from the ground upwards being the predictors and the distance to the cut surface as the response variable also showed a significant relationship. The coefficients of all parameters, minimum diameter, average root angle and intercept were significantly different from zero (Pr > |t| < 0.0001). The model is (Equation (2)):

$$Distance\ to\ cut\ surface\ (mm) = 284.51 + 0.172 \times minimum\ diameter\ (mm) - 3.997 \times root\ angle\ (°) \qquad (2)$$

An increasing diameter as well as a decreasing root angle as the distance from the cut surface downwards increased was plausible. The adjusted R-square was 0.61 and linearity between the predictor variables and the distance to the cut surface existed. The Shapiro-Wilk W test provided a *p*-value of 0.0640, thus, still indicating a normal distribution of the residuals. The variance inflation factor of 2.1 indicated no collinearity of the parameters, whereas the White test indicated that heteroscedasticity was highly significant (Pr > ChiSq is < 0.0001) (Figure 9.) There is a sharp edge at the lower end of both axes.

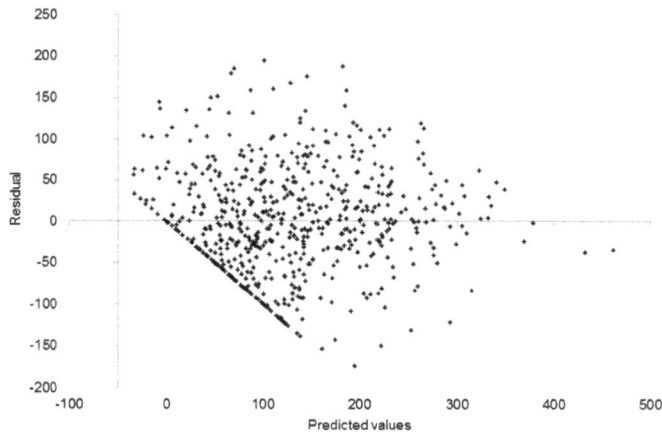

Figure 9. The residuals plotted against the predicted values.

The quantile regression for the 10%-quantile of the data delivered the following coefficients (Equation (3)):

$$\text{Distance to cut surface (mm)} = 124.1323 + 0.2112 \times \text{minimum diameter (mm)} - 3.0189 \times \text{root angle (°)} \quad (3)$$

All coefficients were significantly different from zero. Figure 10 shows the plane spanned by this equation. The border between the green and purple area identified the threshold of the minimum diameter and the average root angles at the cut surface and thus the theoretical threshold for minimum stump height.

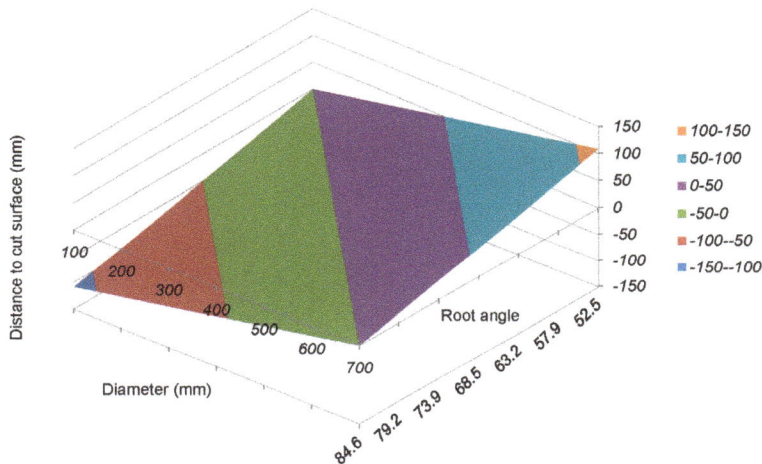

Figure 10. The plane of the 10%-quantile presented in a three-dimensional diagram (stump diameter, root angle, and the distance to cut surface).

Table 1 lists the threshold of diameters and root angles at a distance of zero to the cut surface. For a given diameter, the root angle at the cut surface could be steeper, but not lower than shown in Table 1. Conversely, for a given root angle the minimum diameter could be smaller, but not larger than listed in the table. Only 16% of the stumps had a root angle at the cut surface steeper than 79°. Only two stumps of the subsample exceeded a minimum diameter of 600 mm. No stump was located beyond the threshold of both, 600 mm and 79.2°. This corresponded well to the technical configuration of the harvesting head studied. The opening width of the feed rollers was 550 mm, thus indicating the limit for gripping a tree.

Table 1. The threshold of diameters and root angles at the cut surface.

Minimum Diameter (mm)	100	200	300	400	500	600	700
Root Angle (°)	52.5	57.9	63.2	68.5	73.9	79.2	84.6

Figure 11 shows the frequency of stumps at different length classes indicating the length of stump sections being higher than the 10%-quantile. This analysis implied that 62% of the stumps could have theoretically been cut lower to the ground. The difference between the 10%-quantile and the actual height of the stump could be used as an indicator of unexploited volume since in theory, cutting the tree lower (thus providing a lower stump height) would entail a longer stem than if the tree was cut at a higher position. The length of this potentially unexploited stem wood was mostly shorter than 16.0 cm with an overall average of 9.9 cm corresponding to a total volume of 8.5 dm^3. Considering all stumps, and applying the results from the 10%-quantile analysis, about 0.4% of the core wood of all removed trees was probably unexploited during the mechanized operations.

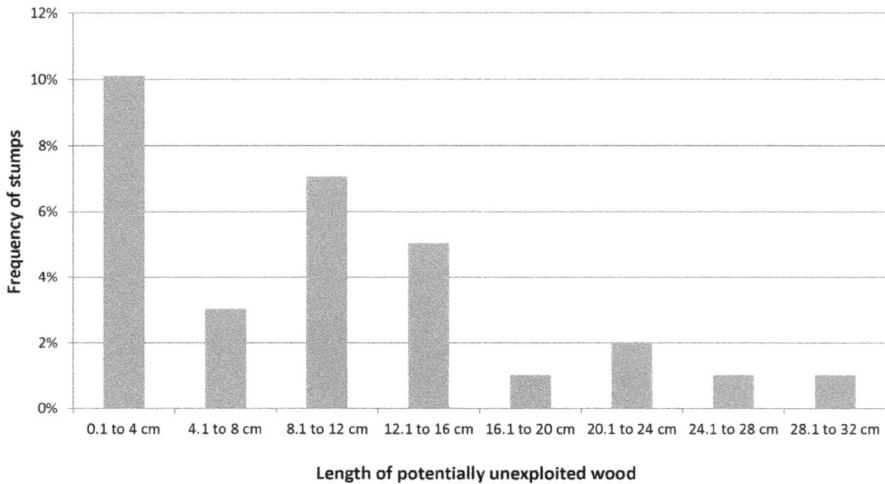

Figure 11. The frequency of stumps presented in classes referring to the length of potentially exploited wood derived by the 10%-quantile analysis.

4. Discussion

4.1. Stump Characteristics

The dataset of 202 stumps supported a positive trend of increasing stump height with an increase in stump diameter. This implies that for those stumps that were close to the maximum operating capability of the harvesting head, stump diameter should have an impact on stump height. Despite the relatively small stump sizes in relation to the maximum opening of the harvesting head, high variation in stump height was observed suggesting that the height at which a stump is cut during mechanized harvesting operations might not be entirely related to physical parameters of the stump or the maximum opening of the harvesting head.

4.2. Root Flares and Root Collar

A positive linear trend could be noticed between the number of root flares per stump and stump height. A much stronger relationship was apparent between the number of root flares per stump and stump diameter. The trend through the entire dataset can be explained as large diameter trees often grow supporting roots to stabilize themselves against side-pressure associated with wind forces. All stumps with no visible above-ground root flares in the dataset had, therefore, relatively small diameters. The development of the trend implied that, when a certain above-ground root flare mass was grown, this effect was getting smaller. By comparing the diameter segments, a second trend could be seen. The point clouds were moving to the top right corner with an increase in diameter class. This also seemed to be caused by the increasing diameter and not by flatter angles. The initial trend was similar throughout the diameter classes but was less apparent at higher diameters.

4.3. Machine Operating Trail

At the onset of the study, we anticipated that as the distance from a stump to the machine operating trail increased, that the operator visibility would be hindered and the lifting force of the boom would decrease, both contributing to higher stumps. However, the horizontal distance between sampled stumps and the machine operating trail did not influence diameter and height of the stumps that were scanned. Of course, the diameter of a tree influences its weight, which can become a limiting factor in relation to the maximum reach of the harvester and its ability to perform adequate and safe felling, but for the study, a consistent deviation of all stump diameters and heights up to the maximum reach of the boom was reported. For some instances, distances measured between the center of the machine operating trail and the center of the sample stumps was greater than the reach of the harvester boom. It is possible that on some occasions, the harvester operator performed so-called poke-ins or pockets with the harvester to be able to reach trees located further away from the trail. With the presence of brush mats coupled with a single machine pass, these pockets can be difficult to detect in the scan. The likelihood of this scenario occurring was increased since operations were performed following a wind-throw. In such conditions, all trees were felled and processed with machines to ensure the safety of all workers. Second, it is also possible that the bird's eye view projection of the machine operating trail centerline was not located precisely because of too high and dense vegetation on the trail at the end of the field campaign. Therefore, in further projects it is recommended to scan early at the beginning of the growing season or control competitive vegetation such as Himalayan balsam (*Impatiens glandulifera* Royle). This particular vegetation became a significant problem by covering up the operating trails and masking them in the scans.

4.4. Concurrent Effects on Stump Height

The diameter of a stump and the shape of the root collar at the cut surface together had a significant effect on stump height. The model explained even half of the variation of stump heights. The minimum diameter contributed more to the explanation of stump height variation than the average or the maximum diameter. This was plausible because it was rather the minimum diameter which limited the grabbing of the tree by the harvesting head, since the later could be swiveled during the positioning at the base of a tree.

Somewhat contradictory appeared the observation that the stump height increased with the diameter, as well as the number of root flares, but in the multiple regression model the number of root flares revealed no significance. The regression model showed no critical multicollinearity of the diameter and the number of root flares when both parameters were included into the model. But in a model with the predictors "root angle" and "number of root flares", the coefficient of the root number also became significantly different from zero. However, this model without the predictor "minimum diameter" had an adjusted R-square of only 0.14. Thus, the number of root flares other than the diameter could explain hardly any variation of the stump heights.

The minimum diameters and average root angles at all 5 cm-segments from the ground upwards seemed to have a significant relation to their distance to the cut surface. However, this model violated an important assumption of linear regression models, because the residuals were not homogenous as expressed by the scatterplot of the residuals showing a sharp edge. Nevertheless, this edge was artificial, because no negative distances to the cut surface were allowed. The edge was marked by the straight line where the negative residuals equaled the predicted values. At a predicted distance of e.g., 100 mm to the cut surface the minimum residual that could occur was −100 mm. The deviation to the negative side could not be greater, because no negative distances to the cut surface were in the data set. If we had measured the tree before logging, we could have determined the distances from above the cut surface too and these distances could have been introduced as negative values into the model. Thus, we would have received a homogenous distribution of the residuals. There might also be certain physical factors located higher than the cut surface that could have influenced stump height but were not considered in the study since only the stump sections were scanned. Further investigations should also consider the shape of the tree before harvesting via a scan of the first 2–3 meters from the ground surface.

A quantile regression applied to the dataset of the last model revealed the lower limit of stump heights, which was imposed by the diameter and the root collar. The model disclosed a limit of minimum diameters at the cut surface of 600 mm. The limiting factor of the particular harvesting head was the opening width of the feed-rollers, which was 550 mm. However, this point was not detected as it was located on the log rather than on the measured stump. The center of the feed-rollers was at a distance of approximately 0.75 m from the level of the saw bar. Therefore, a diameter at the cut surface slightly wider than the maximum opening of the feeding rollers was plausible.

Almost one-third of all stumps could have been cut at a lower level according to the observed limit. There might have been reasonable explanations for cutting the trees at higher levels but they could not be assigned to the size of the stump and the root collar. In the study, the amount of wood which was not exploited was low (0.4% of core wood) but could easily increase in the case of higher stumps. We do not know the variation between different logging operations. Perhaps the stump heights were close to the optimum in our case study and our case is far away from the average. Considering a unit price of 100 €/m^3 for Norway spruce sawlog of high quality and an average tree diameter of 1 m^3, a monetary loss of about 0.40 €/m^3 could be anticipated. At first sight, this might seem quite trivial but when considering that about 60% of the harvests on public forests (approx. 3,000,000 m^3) in Bavaria are performed with fully-mechanized systems, the potential value loss could be considerable.

4.5. Review of the Applied Methods

The thresholds detected in this study cannot be generalized as they refer to the specific configuration of the harvesting head. A harvesting head with a greater opening width of the feed-rollers should allow larger diameters at the cut surface. In other cases, the limiting factor might be the width of the lower knifes of the harvesting head. Further studies with different harvesting heads could disclose the effect of the configuration on the cutting level. We recommend scanning the trees before logging and the stumps after logging. Through this method, the shape of the stem above the cutting level could also be included in the model and the distribution of the residuals should be homogenous. Additional attributes of a tree above the cut surface could also be included in the explanatory model.

Within this study, the operator's line-of-sight and visibility within the harvester cabin was not considered as the field measurement campaign was performed after the completion of the operations. Nevertheless, as one scan was always taken from the intended machine operating trail (clear visibility towards the stump was indispensable), a clear line-of-sight was present for all measured stumps. In a further approach, recording the angle formed between the boom being extended from the machine towards the stump could provide additional information on the position of the harvester and the associated visibility of the operator, thus, further helping to understand the effect of visibility on stump height.

Measuring the minimum diameter at grooves of the contour of the cut surface delivered results which certainly do not refer to the configuration of the harvesting head. In this study, minimum diameters were measured like a caliper measures only for the subsample of 100 stumps. Further studies should measure the diameter in this way, too.

Lastly, we could have chosen other predictor variables. We also could have taken the minimum root angle instead of the average. In fact, we tested different models but the average angle of all roots at one layer delivered better results measured by R-square as compared to the minimum value.

5. Conclusions

In total, 202 mechanically harvested stumps were examined through the use of a TLS, which proved to be a valuable tool for data recording and ensuing assessment. The distance between a stump and the nearest machine operating trail showed no influence on stump height nor did it seem to affect stump diameter under the tested conditions of relatively flat terrain. The diameter alone was able to explain about 20% of the variation in stump heights. The number of above-ground root flares per stump had no noticeable influence on stump height but stump diameters were greater with an increase in the number of roots. The diameter of a stump and the shape of the root collar at the cut

Forests **2018**, *9*, 709

surface together had a significant effect on stump height as the model explained half of the variation of stump heights.

Future studies with expanded data sets could shed more light onto the influence of root collar geometry particularly if they are conducted to include physical features of the bottom log on stump height following mechanized forest operations.

Author Contributions: All authors conceived the experiment while E.R.L. and J.B.H. performed the experiments; All authors analyzed the data and contributed to the written manuscript.

Acknowledgments: The authors wish to extend gratitude to Johannes Windisch, Philipp Gloning and Fabian Schulmeyer for providing insight during the conception of the project. We also wish to thank Michael Miesl for his assistance during the field work and Siegfried Waas for preliminary data preparation.

Conflicts of Interest: The authors declare no conflict of interest.

References

1. Bayerische Staatsforsten. *Statistikband*; Bayerische Staatsforsten: Regensburg, Germany, 2017; p. 30.
2. Gardiner, B.; Blennow, K.; Carnus, JM.; Fleischer, P.; Ingemarson, F.; Landmann, G.; Lindner, M.; Marzano, M.; Nicoll, B.; Orazio, C.; et al. *Destructive Storms in European Forests: Past and Forthcoming Impacts*; Final Report to European Commission–DG Environment; European Forest Institute: Joensuu, Finland, 2013; p. 139.
3. Nick, L. Forstmaschinenstatistik 2007–noch einmal ein Erfolgsjahr. *Forsttech. Inf.* **2008**, *11*, 117–119.
4. Visser, R.; Stampfer, K. Expanding ground-based harvesting onto steep terrain: A review. *Croat. J. For. Eng.* **2015**, *2*, 321–331.
5. Food and Agriculture Organization of the United Nations. Code of Practice for Forest Harvesting in Asia-Pacific. Pre-Harvest Field Preparation. Available online: http://www.fao.org/docrep/004/AC142E/ac142e0d.htm (accessed on 21 November 2016).
6. Hilker, T.; van Leeuwen, M.; Coops, N.C.; Wulder, M.A.; Newnham, G.J.; Jupp, D.L.B.; Culvenor, D.S. Comparing canopy metrics derived from terrestrial and airborne laser scanning in a Douglas-fir dominated forest stand. *Trees* **2010**, *24*, 819–832. [CrossRef]
7. Moorthy, I.; Miller, J.R.; Hu, B.; Chen, J.; Li, Q. Retrieving crown leaf area index from an individual tree using ground based lidar data. *Can. J. For. Res.* **2008**, *34*, 320–332.
8. Strahler, A.H.; Jupp, D.L.B.; Woodcock, C.E.; Schaaf, C.B. Retrieval of forest structural parameters using a ground based lidar instrument. *Can. J. For. Res.* **2008**, *34*, 426–440. [CrossRef]
9. Bao, Y.; Ni, W.; Wang, D.; Yue, C.; He, H.; Verbeeck, H. Effects of tree trunks on estimation of clumping index and LAI from HemiView and Terrestrial LiDAR. *Forests* **2018**, *9*, 144. [CrossRef]
10. Hopkinson, C.; Chasmer, L.; Young-Pow, C.; Treitz, P. Assessing forest metrics with a ground-based scanning lidar. *Can. J. For. Res.* **2004**, *34*, 573–583. [CrossRef]
11. Watt, P.J.; Donoghue, D.N.M. Measuring forest structure with terrestrial laser scanning. *Int. J. Rem. Sens.* **2005**, *26*, 1437–1446. [CrossRef]
12. Henning, J.; Radtke, P. Detailed stem measurements of standing trees from ground-based scanning Lidar. *For. Sci.* **2006**, *52*, 67–80.
13. Maas, H.G.; Bienert, A.; Scheller, S.; Keane, E. Automatic forest inventory parameter determination from terrestrial laser scanner data. *Int. J. Rem. Sens.* **2008**, *29*, 1579–1593. [CrossRef]
14. Yu, X.; Liang, X.; Hyyppä, J.; Kankare, V.; Vastaranta, M.; Holopainen, M. Stem biomass estimation based on stem reconstruction from terrestrial laser scanning point clouds. *Rem. Sens. Let.* **2013**, *4*, 344–353. [CrossRef]
15. Yan, Y.; Xia, M.; fan, S.; Zhan, M.; Guan, F. Detecting the competition between Moso bamboos and broad-leaved trees in mixed forests using a terrestrial laser scanner. *Forests* **2018**, *9*, 520. [CrossRef]
16. Bayer, D.; Siefert, S.; Pretzsch, H. Structural crown properties of Norway spruce (*Picea abies* L. Karst.) and European beech (*Fagus sylvatica* L.) in mixed versus pure stands revealed by terrestrial laser scanning. *Trees* **2013**, *27*, 1035–1047. [CrossRef]
17. Bayer, D.; Pretzsch, H. Reactions to gap emergence: Norway spruce increases growth while European beech features horizontal space occupation—Evidence by repeated 3D TLS measurements. *Silva Fennica* **2017**, *51*, 20. [CrossRef]

18. Beyer, R.; Bayer, D.; Letort, V.; Pretzsch, H. Validation of a functional-structural tree model using terrestrial Lidar data. *Ecol. Model.* **2017**, *2017. 357*, 55–57. [CrossRef]
19. Pyörälä, J.; Kankare, V.; Rikala, J.; Holopainen, M.; Sipi, M.; Hyyppä, J.; Uusitalo, J. Comparison of terrestrial laser scanning and X-ray scanning in measuring Scots pine (*Pinus sylvestris*) branch structure. *Int. J. For. Eng.* **2018**, *33*, 291–298. [CrossRef]
20. Liski, J.; Kaasalainen, S.; Raumonen, P.; Akujärvi, A.; Krooks, A.; Repo, A.; Kaasalainen, M. Indirect emissions of forest bioenergy. Detailed modeling of stump-root systems. *GCB Bioenergy* **2014**, *6*, 777–784. [CrossRef]
21. Komatsu Forest. 360.2 Spezifikationen. 2015. Available online: http://www.komatsuforest.de/default.aspx?id=10778&mode=specs&rootID=&productId=22.07.2015 (accessed on 24 May 2018).
22. FARO Technologies Inc. *FARO Laser Scanner Focus3D User Guide*; FARO: Lake Mary, FL, USA, 2015; p. 186.
23. Han, H.-S.; Renzie, C. Effect of ground slope, stump diameter, and species on stump height for feller-buncher and chainsaw felling. *Int. J. For. Eng.* **2005**, *16*, 81–88. [CrossRef]
24. Ministry of Forests, Lands and NRO. *Cruising Procedures Manual*; British Columbia: Victoria, Canada, 2014; Volume 6. Available online: https://www.for.gov.bc.ca/ftp/hva/external/!publish/web/manuals/cruising/chapters/Ch6.pdf (accessed on 21 November 2016).
25. UCLA: Statistical Consulting Group. Introduction to SAS. 2017. Available online: https://stats.idre.ucla.edu/sas/webbooks/reg/chapter2/regressionwith-saschapter-2-regression-diagnostics/ (accessed on 28 May 2018).
26. Colin, C. *An Introduction to Quantile Regression and the QUANTREG Procedure*; Nelson, G.S., Ed.; 2005; pp. 213–230. Available online: http://www2.sas.com/proceedings/sugi30/toc.html (accessed on 20 June 2018).

forests

MDPI

Article

Sensitivity of Codispersion to Noise and Error in Ecological and Environmental Data

Ronny Vallejos [1,*], **Hannah Buckley** [2], **Bradley Case** [2], **Jonathan Acosta** [3] **and Aaron M. Ellison** [4]

1   Departamento de Matemática, Universidad Técnica Federico Santa María, Avenida España 1680, Valparaíso 2340000, Chile
2   School of Science, Auckland University of Technology, 55 Wellesley Street East, Auckland 1010, New Zealand; hannah.buckley@aut.ac.nz (H.B.); bradley.case@aut.ac.nz (B.C.)
3   Instituto de Estadística, Pontificia Universidad Católica de Valparaíso, Av. Brasil 2950, Valparaíso 2340000, Chile; jonathan.acosta@pucv.cl
4   Harvard Forest, Harvard University, 324 North Main Street, Petersham, MA 01366, USA; aellison@fas.harvard.edu
*   Correspondence: ronny.vallejos@usm.cl; Tel.: +56-32-2654964

Received: 26 September 2018; Accepted: 21 October 2018; Published: 29 October 2018

check for updates

Abstract: Understanding relationships among tree species, or between tree diversity, distribution, and underlying environmental gradients, is a central concern for forest ecologists, managers, and management agencies. The spatial processes underlying observed spatial patterns of trees or edaphic variables often are complex and violate two fundamental assumptions—isotropy and stationarity—of spatial statistics. Codispersion analysis is a new statistical method developed to assess spatial covariation between two spatial processes that may not be isotropic or stationary. Its application to data from large forest plots has provided new insights into mechanisms underlying observed patterns of species distributions and the relationship between individual species and underlying edaphic and topographic gradients. However, these data are not collected without error, and the performance of the codispersion coefficient when there is noise or measurement error ("contamination") in the data heretofore has been addressed only theoretically. Here, we use Monte Carlo simulations and real datasets to investigate the sensitivity of codispersion to four types of contamination commonly seen in many forest datasets. Three of these involved comparing codispersion of a spatial dataset with a contaminated version of itself. The fourth examined differences in codispersion between tree species and soil variables, where the estimates of soil characteristics were based on complete or thinned datasets. In all cases, we found that estimates of codispersion were robust when contamination was relatively low (<15%), but were sensitive to larger percentages of contamination. We also present a useful method for imputing missing spatial data and discuss several aspects of the codispersion coefficient when applied to noisy data to gain more insight about the performance of codispersion in practice.

Keywords: codispersion coefficient; codispersion map; imputation; kriging; measurement error; missing observations; spatial noise

---

1. Introduction

Spatial associations are a fundamental aspect of most ecological and environmental data, including size-frequency distributions of trees, their co-occurrence and diversity, and the relationships between trees and underlying edaphic characteristics or topographic variables such as distance to water table, slope, and aspect. Although accounting for spatial covariation has become routine in ecological data analysis [1], forest ecologists have been slower to appreciate and account for anisotropic patterns and processes (but see, e.g., [2]). Codispersion [3] measures lag-dependent spatial covariation in two or more spatial processes, which may be anisotropic. Codispersion recently has been used to examine interactions between species, and relationships between species distributions and underlying environmental gradients in large (>25-ha) forest dynamics plots. These analyses have provided new insights into potential ecological processes that underlie observed patterns in co-occurrence between pairs of tree species [4], in relationships between attributes of individual tree species and underlying edaphic characteristics [5], and in forest structure through time [6].

All applications of codispersion analysis that have been published to date have assumed either that there are no errors in the datasets or that any errors that are present would have no effect on the analysis. These assumptions are clearly unrealistic. The goal of this paper is to better understand how sensitive the codispersion coefficient is to different types of noise and measurement error ("contamination") in analyzed datasets, and address the implication of this sensitivity for spatial analysis of data collected on forest structure and composition. We approach this goal by using Monte Carlo simulation studies to examine several classes of noise that we would expect to occur in datasets or images analyzed using codispersion. Our focus here is on the analysis of data gathered from large forest plots either by remote sensing or on-the-ground sampling that are used to describe forest structure or test hypotheses regarding the relationship between spatial distribution of trees and edaphic variables. However, the results are generalizable to any dataset for which there is either measurement error in the spatial data collected or process error in the spatial models that are used in subsequent analysis.

Measurement (observation) error can occur in several ways. For example, trees can be misidentified by observers or pixels in remotely-sensed images can be misclassified. Either can affect inferences about interspecific relationships or associations between plant species and edaphic characteristics. To examine the effect of these "simple" observation errors (misidentification or misspecification) on codispersion analysis, we added statistical noise to a fixed number of random points (or pixels) in a dataset (here, a remotely-sensed image of a forest stand) either as white noise (spatially independent and identically distributed) or as a spatially-dependent process.

Another type of measurement error would be when clusters of individuals are missed or overlooked in a census of a forest stand, either through human error of if clusters of pixels in a remotely-sensed image are unmeasurable because of cloud cover. Spatial analysis of such data would require that these gaps be filled, and we present and assess the consequences of different algorithms for interpolation prior to calculation of codispersion [7].

The flip-side of interpolating missing data is to smooth sparsely-collected data (e.g., soils data smoothed using kriging, splines, etc.); the smoothed surface is subsequently sampled at specific (otherwise unsampled) points to test for associations between individual tree species and (estimated) local environmental (e.g., soil) properties. Errors here can occur because of mis-specified models or because too few data are available to construct a reliable smoothed surface. We examined these issues for the assessment of relationships between trees and soil characteristics when different amounts of modeled error were introduced into the environmental data as a result of kriging surfaces derived from complete or "thinned" datasets; the latter mimicked datasets with missing values.

In all of these cases, the effects of error were tested in one of two ways. For single datasets (images) to which we added random or clustered errors, we calculated the codispersion between the original dataset (image) and the contaminated version of itself. In those cases, we examined differences between high- and lower-quality (error-filled) images. For species–soil relationships, we compared the results obtained using kriged surfaces derived from the complete sample of soil properties versus those derived from soil samples missing random points (the "thinned") data.

In Section 2, we describe the codispersion coefficient and a way to visualize it (i.e., a codispersion map). We also specify the different types of contamination and observation error that we added to both real and simulated datasets, and describe the method we used for imputing missing data. Results are presented in Section 3, and discussed in Section 4. Technical details on simulation and imputation algorithms are given in Appendix A.

2. Methods

2.1. Preliminaries and Notation

Here, we briefly introduce the notion of the codispersion coefficient, first described in [8], and the graphical codispersion "map" developed in [3] and applied to forest data in [4]. These statistical entities characterize the spatial correlation between two spatial processes as a function of the separation distance (lag) between the points.

Let us consider two spatial processes $\{X(s) : s \in D \subset R^2\}$ and $\{Y(s) : s \in D \subset R^2\}$, where both processes are defined on a part of a region $D \subset R^2$ (or on a rectangular lattice $D \subset Z^2$). For two intrinsically stationary processes $X(\cdot)$ and $Y(\cdot)$, the codispersion coefficient is defined as

$$\rho_{XY}(h) = \frac{\mathbb{E}\left[(X(s+h) - X(s))(Y(s+h) - Y(s))\right]}{\sqrt{\mathbb{E}[X(s+h) - X(s)]^2 \mathbb{E}[Y(s+h) - Y(s)]^2}}, \tag{1}$$

for $h = (h_1, h_2)$, $s + h \in D$. This coefficient shares several properties with Pearson's correlation coefficient (r). First, the structure of ρ_{xy} is computationally similar to r. Second, like r, $-1 \leq \rho_{XY}(h) \leq 1$, which facilitates its interpretation because the upper and lower bounds define perfect negative or positive spatial association, respectively. Unlike r, however, ρ_{xy} depends on the spatial lag h, which emphasizes that spatial correlation is a value associated with a distance on the plane. This facilitates the computation of correlation for different distances and directions on the space. In this sense, Pearson's correlation is a crude measure of the spatial association between two processes.

For n sampling sites $s_1, \ldots, s_n$, the sample-based estimator of (1), based on the method of moments, is

$$\widehat{\rho}_{XY}(h) = \frac{\sum\limits_{s \in N(h)} (X(s) - X(s+h))(Y(s) - Y(s+h))}{\sqrt{\sum\limits_{s \in N(h)} (X(s) - X(s+h))^2 \sum\limits_{s \in N(h)} (Y(s) - Y(s+h))^2}}, \tag{2}$$

where $N(h) = \{(s_i, s_j) : ||s_i - s_j|| \in T(h), 1 \leq i, j \leq n\}$, $T(h)$ is a tolerance region around h, and $|| \cdot ||$ denotes the Euclidean norm in $\mathbb{R}^2$. The estimator of the codispersion coefficient given in Equation (2) can be computed for any fixed spatial lag h. This computation can be difficult if the number of points is small or if $N(h)$ is an empty set. We emphasize that the empirical estimator of the codispersion coefficient makes real sense when the processes are defined on a finite rectangular grid in a two-dimensional space that corresponds to the assessment of the similarities between two digital images [9].

When the codispersion coefficient is computed for many directions, it is useful to display those values on a single graph. Vallejos [3] suggested a graphical tool called the codispersion map to visualize the spatial correlation between two sequences on a plane. The estimated values illustrated by the codispersion map are based on Equation (2). A finite grid on the plane is first defined on which the codispersion coefficient will be computed for each location in that grid. The codispersion map itself is the graph of $\widehat{\rho}_{XY}(h_1, h_2)$ versus (h_1, h_2); plotting the codispersion map summarizes the information about the spatial association between two sequences in a radial way on the plane circumscribing the map in a semicircle of fixed radius.

Note that $\widehat{\rho}_{XY}(h)$ does not capture similarity that is related to the patterns or shapes that are present in the images. Rather, it captures the spatial dependence between the processes for a given lag distance h.

2.2. Types of Error

In spatial modeling and time series, several types of error (*a.k.a.* noise) can be specified [10,11]. Here, we considered five types of error frequently observed in spatial data; our examples are drawn from data collected from forest stands, which to date have been the primary testbed for ecological applications of codispersion analysis.

1. **"Salt-and-pepper" noise on an image**: Salt-and-pepper noise—so-called because of its resemblance to dust on images that appears to have been distributed by a salt or pepper shaker—is used widely in image processing and computational statistics to represent real distortions [12] and to generate different scenarios via Monte Carlo simulation [13]. Salt-and-pepper noise can be added to an image using a simple algorithm:

   Assume that $\{X(s) : s = (i, j), 1 \leq j \leq m, 1 \leq j \leq n\}$ is the original image whose individual observations are points or pixels representing leaves or trees and $Y(i, j)$ is the contaminated image with salt-and-pepper noise such that the additive noise is drawn from a normal distribution with mean = 0 and variance τ^2, with $\tau^2 \gg \sigma^2$, where σ^2 is the variance of $X(i, j)$. The contamination is located randomly in space such that a small percentage of observations are corrupted with a probability δ [3]. Specifically,

   $$Y(i,j) = X(i,j) + v(i,j)V(i,j), \; i = 1, \ldots, m, j = 1, \ldots, n, \tag{3}$$

   where the $v(i,j)V(i,j)$ is an outlier generating process such that $v(i,j)$ is a zero-one process with $\mathbb{P}[v(i,j) = 1] = \delta$ and $\mathbb{P}[v(i,j) = 0] = 1 - \delta$, and $V(i,j) \sim \mathcal{N}(0, \tau^2)$.

   We used Monte Carlo simulations of (3) to generate salt-and-pepper noise on a 5616×3744-pixel aerial image of a forest stand at Harvard Forest in Petersham, MA, USA (Figure 1). We considered $\sigma^2 = 1$, $\tau^2 = \{1, 5, 10\}$, and the percentage of contamination $\delta = \{0.05, 0.1, 0.25\}$. We conjectured that the codispersion coefficient would be robust for $\delta \leq 0.05$. That is, for relatively small amounts of measurement error, we could still recover the relevant spatial information present in the remotely-sensed image.

   In Figure 1a, we illustrate the noise-free image. Figure 1c,e,g is contaminated versions of the original one when $\delta = \{0.05, 0.10, 0.25\}$. The corresponding perspective plots shown in Figure 1b,d,f,h depict the effect of contamination on the gray intensities. The greater the contamination, the greater the dispersion, which is plotted on the z-axis of the three-dimensional scatter plots displayed in Figure 1.

   We then compared the codispersion calculated for the original image to that calculated for the contaminated images. In addition to the reference image shown in Figure 1a, we considered other aerial images. The codispersion maps of these images are presented in the supplementary material for this paper. We emphasize that the computation of the codispersion coefficient requires that both processes are measured over the same domain, thus the codispersion between a reference image and its contaminated versions make sense. To address the codispersion between two images taken from different scenarios (for instance, images displayed in the supplementary material), rasterized versions of the original images could be considered following the guidelines given in [4].

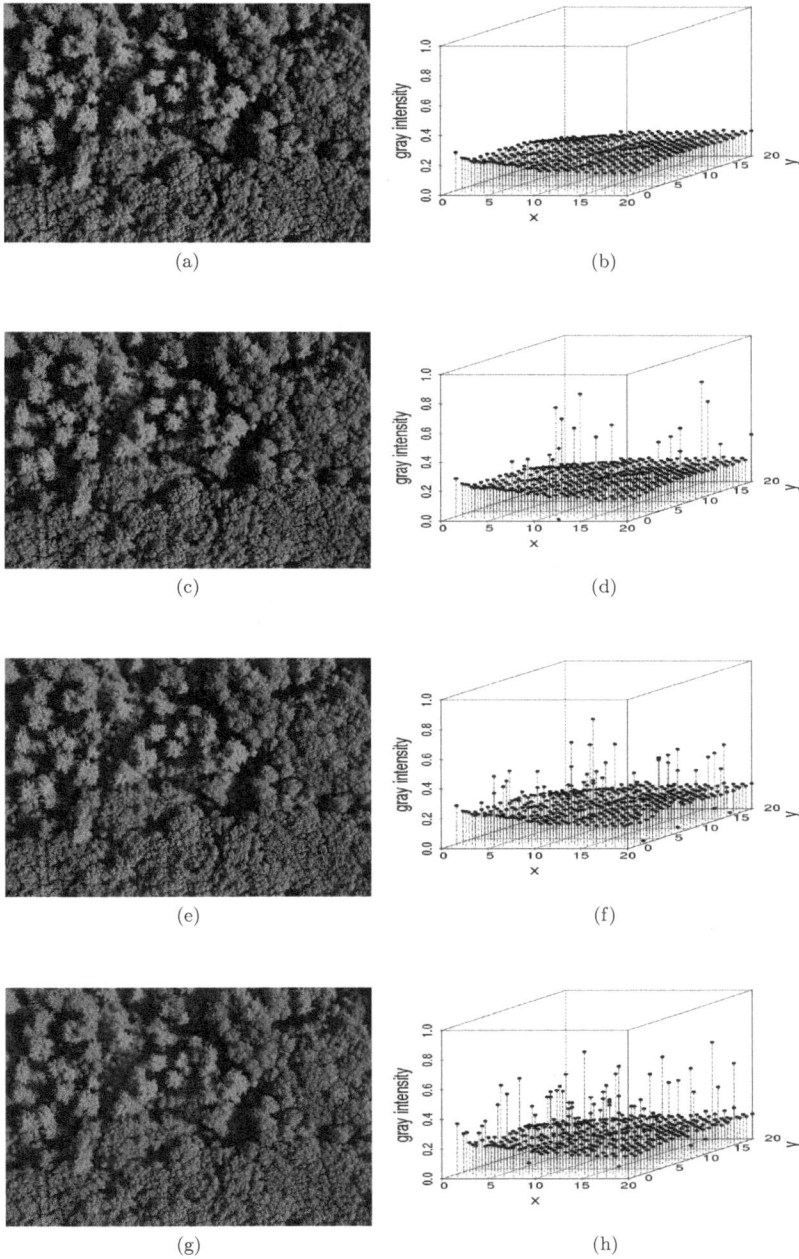

Figure 1. (**a**) reference image of size 5616 × 3744 pixels taken above a section of forest at the Harvard Forest, Petersham, MA, USA and (**b**) its corresponding gray scale values. (**c,e,g**) are the same image distorted with increasing amounts of salt-and-pepper noise. The percentages of contamination are 5%, 10%, and 25%, respectively; (**d,f,h**) show the change in gray intensity after the addition of salt-and-pepper noise to the images.

2. **Salt-and-pepper noise on dependent processes**: More generally, Reference [14] extended the well known Matérn class of covariance functions to a multivariate random field. For multivariate Gaussian and second-order processes, the multivariate Matérn covariance function is defined as

$$M(\boldsymbol{h}|\nu, a) = \frac{2^{1-\nu}}{\Gamma(\nu)}(a||\boldsymbol{h}||)^{\nu} K_{\nu}(a||\boldsymbol{h}||), \tag{4}$$

where $||\boldsymbol{h}||$ is the distance lag, K_{ν} is a modified Bessel function of the second kind, $a > 0$ is a spatial scale parameter, and $\nu > 0$ is a smoothness parameters that defines the Hausdorff dimension and the differentiability of the sample paths. In particular, a Gaussian and second-order stationary process $(X(\boldsymbol{s}), Y(\boldsymbol{s}))^{\top}$, $\boldsymbol{s} \in D \subset \mathbb{R}^2$ has a bivariate Matérn covariance matrix if

$$\begin{pmatrix} C_{11}(\boldsymbol{h}) & C_{12}(\boldsymbol{h}) \\ C_{21}(\boldsymbol{h}) & C_{22}(\boldsymbol{h}) \end{pmatrix}, \tag{5}$$

where $\boldsymbol{h} \in D$, $C_{ii}(\boldsymbol{h}) = \sigma_i^2 M(\boldsymbol{h}|\nu_i, a_i)$ are the marginal covariance functions, with variance parameter $\sigma_i^2 > 0$, smoothness parameter $\nu_i > 0$, and scale parameter $a_i > 0$ for $i = 1, 2$. $C_{12} = C_{21} = \rho_{12}\sigma_1\sigma_2 M(\boldsymbol{h}|\nu_{12}, a_{12})$ is the cross-covariance function, with correlation coefficient ρ_{12}, smoothness parameter ν_{12}, and scale parameter a_{12}. In all cases, $M(\cdot)$ is the function defined in Equation (4). The parsimonious bivariate Matérn model has the restriction

$$|\rho_{12}| \leq \frac{(\nu_1\nu_2)^{1/2}}{\frac{1}{2}(\nu_1 + \nu_2)}. \tag{6}$$

The correlation between the spatial variables $X(\cdot)$ and $Y(\cdot)$ is controlled by the parameter ρ_{12}, which allows one to generate bivariate Gaussian spatial processes with different levels of dependence. The spatial correlation defined by Equation (6) is not necessarily bounded by 1. Without loss of generality, it can be assumed that the mean of the bivariate process is zero, but the theory works well for any bivariate process with mean $(\mu_1, \mu_2)^{\top}$. Any type of contamination can be applied over the generated dependence data. In this case, we applied salt-and-pepper noise.

We generated dependent random fields from the bivariate Matérn class of covariance functions described in Equation (5) by Monte Carlo simulation using the R package RandomFields [15]. We then added the salt-and-pepper noise, varying the additional parameter ρ_{12}, which represents the known correlation between processes $X(\cdot)$ and $Y(\cdot)$.

Figure 2 shows one realization of size 512×512 from a bivariate Gaussian process (images (a) and (b)) with correlation equal to 0.8, and $\nu_1 = \nu_2 = \nu_{12} = 0.5$, $\sigma_1^2 = \sigma_2^2 = 1$, $\mu_1 = \mu_2 = 0.5$ and $a_1 = a_2 = 2/512$. Figure 2c,d,e show versions of (b) contaminated with salt-and-pepper noise with the percentage of contamination equal to 5%, 15%, and 25%, respectively. Because the Gaussian process is stationary, images (a) and (b) look very regular (approximately constant mean and variance), and any correlation between them (if it exists) is difficult to observe in the printed images. Other parameters used in the simulation study are $\nu_1 = \nu_2 = 0.5$, $\nu_{12} = 1.5$, $\sigma_1^2 = \sigma_2^2 = 0.125$, $\mu_1 = \mu_2 = 0.5$, $\rho_{12} = 0.1$ and $a_1 = a_2 = 4/512$. The results are similar to the shown here, but with a codispersion map close to zero.

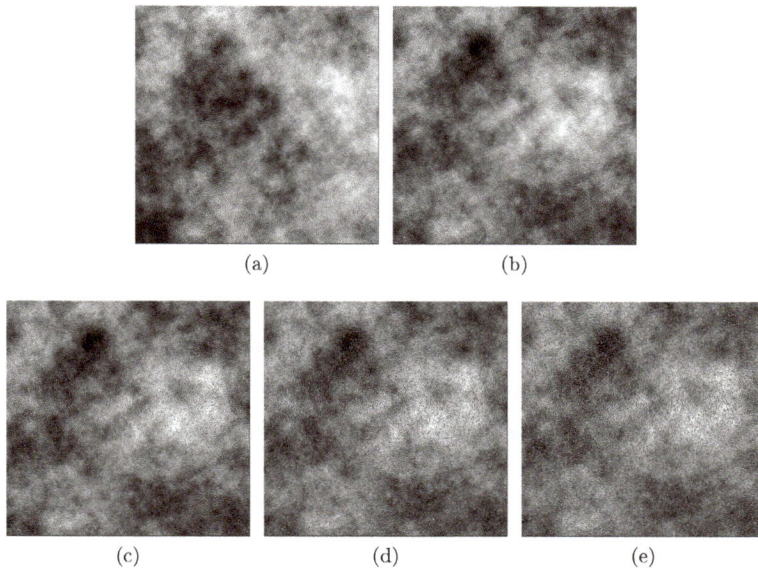

Figure 2. Images (**a**,**b**) are dependent processes generated from a Gaussian process with a covariance matrix as in (5); (**c**–**e**) salt and pepper contamination of (**b**) considering $\delta = (0.05, 0.15, 0.25)$, and $\nu_1 = \nu_2 = \nu_{12} = 0.5$, $\sigma_1^2 = \sigma_2^2 = 1$, $\mu_1 = \mu_2 = 0.5$, $\rho_{12} = 0.8$ and $a_1 = a_2 = 2/N$ in each case, where $N \times N$ is the size of the image.

3. **Missing observations at random locations**: We used the salt-and-pepper scheme to randomly delete n observations. We first defined the percentage of contamination (δ), and then deleted that many observations from the dataset. In practice, we replaced observations with non-observed (NA) at the randomly-selected locations. The main feature of these missing observations is that they are spatially independent of one another, but, for the posterior data analysis, they will remain fixed. The imputation algorithm described in the Appendix A was not applied here because codispersion calculations are not affected when the percentage of contamination δ is small.

In Figure 3, we illustrate the missing-observations-at-random-locations with nine contaminated versions of the original image shown in Figure 1a. The columns show the effect of increasing the percentage of contamination (5%, 15%, and 25%, respectively), and the rows depict the effect of increasing the block size of contaminated pixels, which are 15×15, 30×30, and 60×60 respectively. The contaminated pixels have been colored in white. NAs were ignored in the computation of the codispersion coefficients because for large gaps of missing observations the computation of the codispersion coefficient will be affected for those directions h such that $||h||$ is less than the maximum diameter of the missing block.

Figure 3. Contamination of of the reference image shown in Figure 1a by salt and pepper at random locations. The missing blocks are of size 15×15, 30×30, and 60×60, respectively, shown in the different columns. (**a**–**c**): the proportion of missing blocks is 0.000002; (**d**–**f**): the proportion of missing blocks is 0.000004; (**d**–**f**): the proportion of missing blocks is 0.000008.

4. **Gaps resulting from clusters of missing observations**: Missing values may be clustered, for example, either because of local difficulties in sampling or because large sections of a remotely-sensed image are obscured by, for example, clouds or shadows. We simulated clustered missing observations for the image shown in Figure 1a, given three different pixel sizes for the contaminated block: 200×200, 400×400, and 800×800 (Figure 4). We used simple clustered geometries (squares) for ease of computation. The difference between the previous type of contamination and this one is that, in the former, the contamination consisted of several blocks of small size. Here, we introduced just one gap containing a large number of pixels, which, in Figure 4, is located for illustrative purposes in the center of the image. In our simulations and analysis, the size of the missing block and its location were fixed.

To compute codispersion coefficients for datasets with such large blocks of missing data, we needed to fill the missing gaps (impute missing data) prior to computing the codispersion coefficient. We used and compared two different methods of imputation (gap-filling).

First, the image with a missing gap was represented by a first-order spatial autoregressive process. The fitting of the parameters of the models was done via least-squares estimation following the guidelines given in [16]. This estimation method was studied in [17] and found to yield an approximated image $\widehat{Z}$ of the original one X (see Algorithm 1 in the Appendix A).

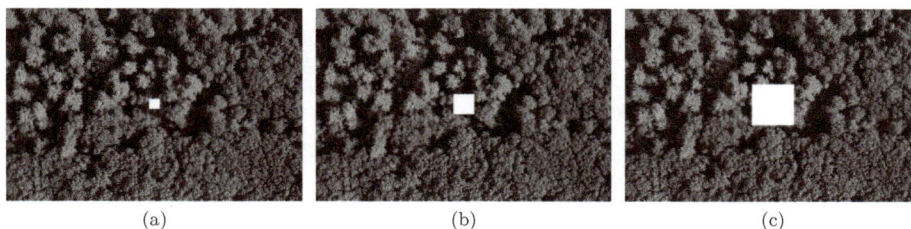

Figure 4. Gaps of missing observations of sizes 200 × 200 (**a**); 400 × 400 (**b**); and 800 × 800 (**c**).

Second, to predict the values of the process in the locations belonging to the missing block, we applied Algorithm 1 to predict missing values in the four closest blocks to the missing gap as is illustrated in Figure A1. This prediction scheme is summarized in Algorithm 2 (Appendix A). Briefly, the first step represents the image intensity by an autoregressive process that assumes that the intensity of any pixel is a weighted average of the intensity of the surrounding pixels. This is a model-based alternative to the average or median commonly computed using the intensities of a moving window across the image. The second step predicts the missing values using similar autoregressive models to represent the surrounding blocks. The predicted value of a pixel belonging to the missing block is a weighted average where the weights are proportional to the distance from the missing pixel to the surrounding blocks.

5. **Sampling error**: Values for edaphic or environmental variables at specific locations in space often are sampled from a smoothed (kriged) surface, which itself was generated from a much smaller set of field observations. The actual information in the kriged surface is a function of both the number of observations and the smoothing parameter of the covariance function [18]. For a pair of spatial point processes $X(\cdot)$ and $Y(\cdot)$ (e.g., individual forest trees and soil nutrient concentrations at each tree, respectively), where the number of observed trees (hundreds to thousands) vastly exceeds the number of soil samples (tens), we kriged the soil chemistry variables after thinning (or not) and then calculated the codispersion between the observed tree diameters and the value of the soil-chemistry variable predicted (at each tree location in $X(\cdot)$) from the kriged surface $\widehat{Y}(\cdot)$ of soil-chemistry data. The kriged surface was computed either from all the data or from "thinned" soil datasets that contained 90% or 80% of the original soil chemistry data [18]. The sampling error here is error in the predicted values at points on the kriged surface caused by fitting the surface to fewer and fewer points in the "thinned" datasets.

To illustrate the effect of this sampling error, we used data from plants and soils collected in the 50-ha forest dynamics plot on Barro Colorado Island, Panamá [19–21]. Of the 299 plant species mapped, identified, and measured every five years in this plot, we used six: *Alseis blackiana, Oenocarpus mapora, Hirtella triandra, Protium tenuifolium, Poulsenia armata, and Guarea guidonia* (Figure 5). The abundances of unique single-stemmed individuals of each of these six species ranged from 993 (*Poulsenia armata*) to 7928 (*Alseis blackiana*), and included species that had a range of positive, negative, and weak associations with measured soil variables [22]. Spatial locations and "diameters at breast height" (at 1.3 m aboveground) of individual trees of each species (excluding dead individuals and individuals with more than one stem) were taken from the seventh (2010) semi-decadal census of the plot.

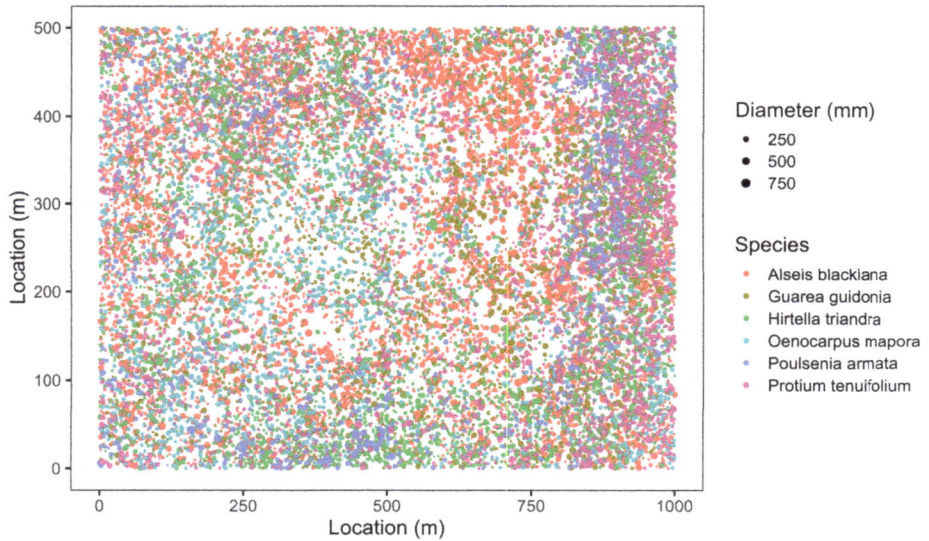

Figure 5. Distribution and size of the six species of trees growing in the 50-hectare plot at Barro Colorado Island, Panamá that we analyzed to assess the effect of sampling error.

Soil samples were collected on a 50-m lattice in 2005 with additional samples taken at finer spatial grains at alternate sampling stations [22]. Soil samples were analyzed for concentrations of 11 elements; we used only data for concentrations of calcium (Ca), phosphorus (P), and aluminium (Al), as these three had the highest loadings on the first three principal axes of a multivariate analysis (NMDS) on the complete soil dataset [22]. We used ordinary kriging in the geoR package [23], version 1.7-5.2, to fit a surface to the data for each soil element and predict its concentration at the location of each tree (Figure 6). Variogram models (exponential, exponential, and wave for Ca, P, and Al, respectively) needed as input for the kriging function were fit to detrended (2nd-order polynomial) data that had been Box–Cox transformed ($\lambda = 0.5$, 1.0, and 1.0 for Ca, P, and Al, respectively); kriging was done on back-transformed data to which the trend had been added. Nuggets were estimated empirically for Ca and P, but the nugget for Al was fixed (following visual inspection of the empirical variogram) equal to 4000. Alternatively, in order to take into account the spatial heterogeneity, one could perform a test to measure the degree of spatial heterogeneity along the lines given in [24], before applying the kriging interpolation.

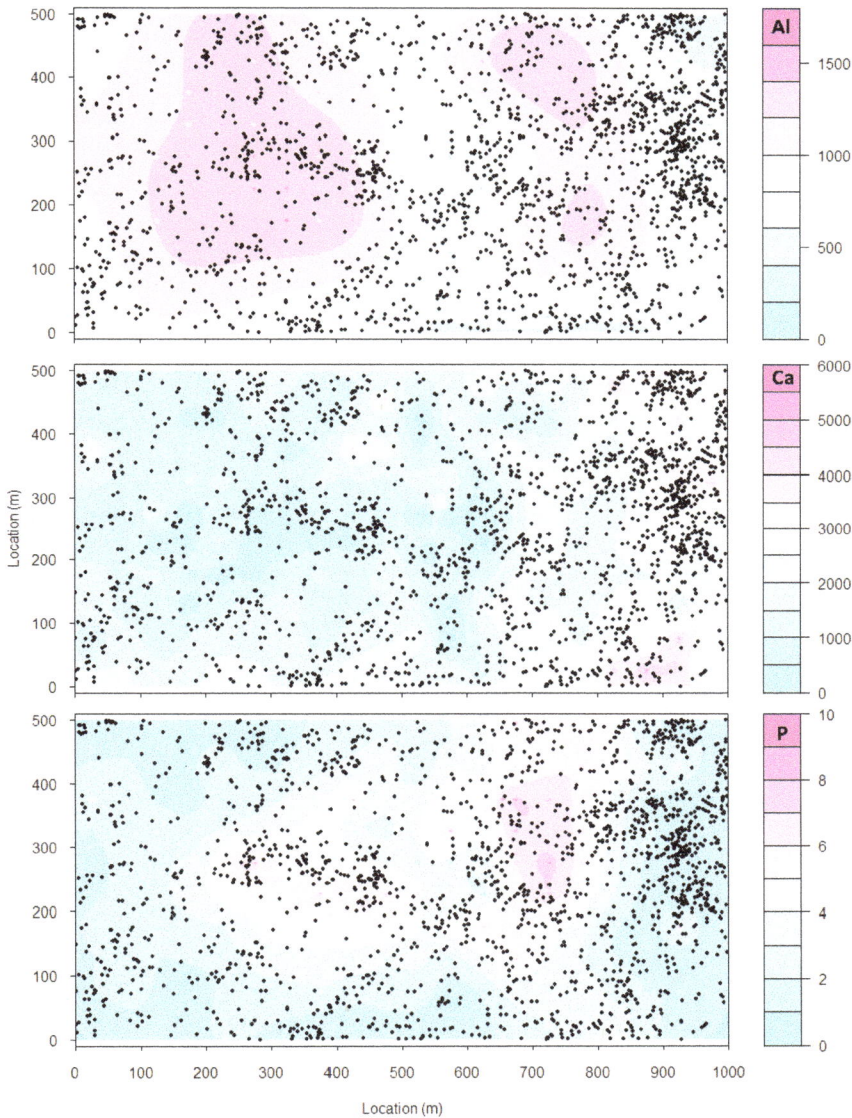

Figure 6. Kriged surfaces of the concentration (mg/kg) of aluminum (Al; top), calcium (Ca; center), and phosphorus (P; bottom) in the 50-hectare plot at Barro Colorado Island, Panamá. Contours were estimated for a regular grid (5-m spacing) based on data from samples taken at approximately 50-m intervals. Interpolated values of mineral concentrations were estimated at individual points (locations of trees) shown on the plots.

3. Results

We used codispersion maps [3,4] to explore the possible patterns and features caused by the introduction of noise and to evaluate the performance of the codispersion coefficient when the process was contaminated with one of the distortions described above. Recall that the generation of the noise is through statistical models that do not necessarily include a particular direction in space. The effects of specific directional contamination on codispersion was investigated in [3].

The only effect observed when the forest image was contaminated with salt-and-pepper noise (Figure 1) was a trend of decreasing codispersion between the original and contaminated images with an increase in the percentage of contamination (in Figure 7, notice a color degradation in the map as the percentage of contamination increases). In the case of the dependent processes generated by a Gaussian process with covariance matrix as in (5), we plotted the codispersion maps between the original and contaminated images displayed in Figure 2. The salt-and-pepper contamination caused a complete loss of correlation between the two images, which were originally correlated strongly ($r = 0.8$; Figure 8). This is in agreement with [3], who reported that visually, it is possible to observe a degradation of the original patterns as an effect of the percentage of contamination. A decrease in codispersion between the original and contaminated images was also observed when noise was introduced through missing observations at random locations or as the missing block size increased (Figure 9).

Figure 4 illustrates how we introduced large gaps of missing values in the center of the reference image shown in Figure 1a. Before computing the codispersion map, we imputed the missing data (Algorithm 2 in the Appendix A). Although the performance of such algorithms strongly depends on the size of the block of missing observations, the construction of it is based on the spatial information contained in the nearest neighbors (Algorithm 1 in the Appendix A). The spatial autoregressive lags in the AR-2D process are fixed when the order of the process is chosen. In this case, three neighbors were considered in a strongly causal set to guarantee an infinite moving average representation of the process. The images filled by the imputation algorithm are shown in Figure 10d–f. The filled areas are smooth in terms of texture and have a smaller variance. Visually, the imputation of the larger missing block looks different from the rest of the image. For small missing blocks, it is difficult to see the imputed values. From Figure 10g–h, we observed that Algorithm 2 was able to recover valuable information and that the codispersion between the original and imputed images in all cases was close to one.

Finally, the codispersion between tree species' diameters (for the six species shown in Figure 5) and the three soil elements (Figure 6) sampled in the Barro Colorado Island plot at three levels of data "thinning" (i.e., estimates of soil properties derived from kriged surfaces of all the soil samples, 90% of them, or 80% of them) showed that codispersion was robust to this form of error. Only the results for the most abundant (Figure 11) and the least abundant (Figure 12) species are shown.

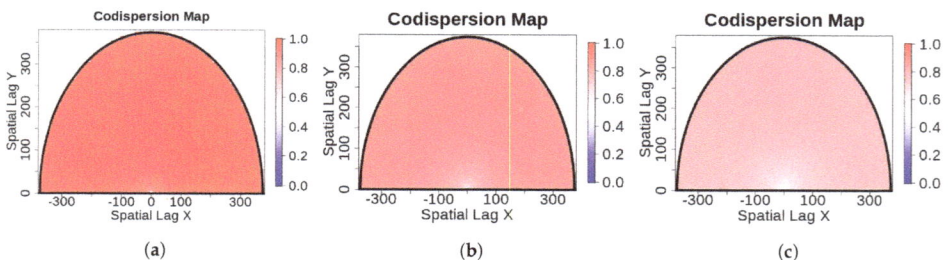

Figure 7. Codispersion map between (a) the original Figure 1a,c (5%); (b) the original Figure 1a,e (15%); (c) the original Figure 1a,g (25%).

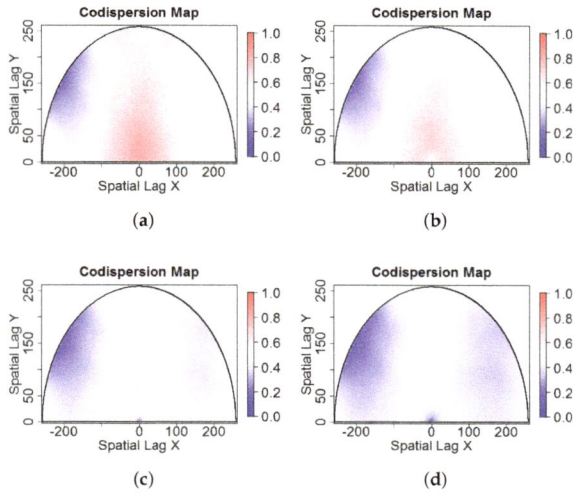

Figure 8. Images (**a–d**) are the corresponding codispersion maps between Figure 2a and the contaminated Figure 2b–e.

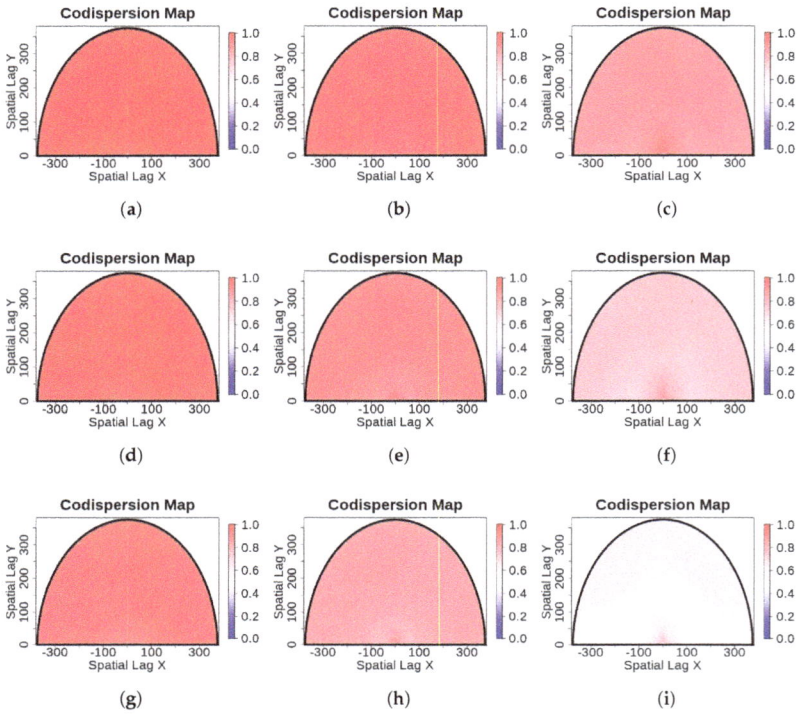

Figure 9. Codispersion map between the reference Figure 1a and the images contaminates with missing observation at random locations depicted in Figure 3a–i. Images (**a**)–(**i**) show how the correlation decreases when the missing block size increases.

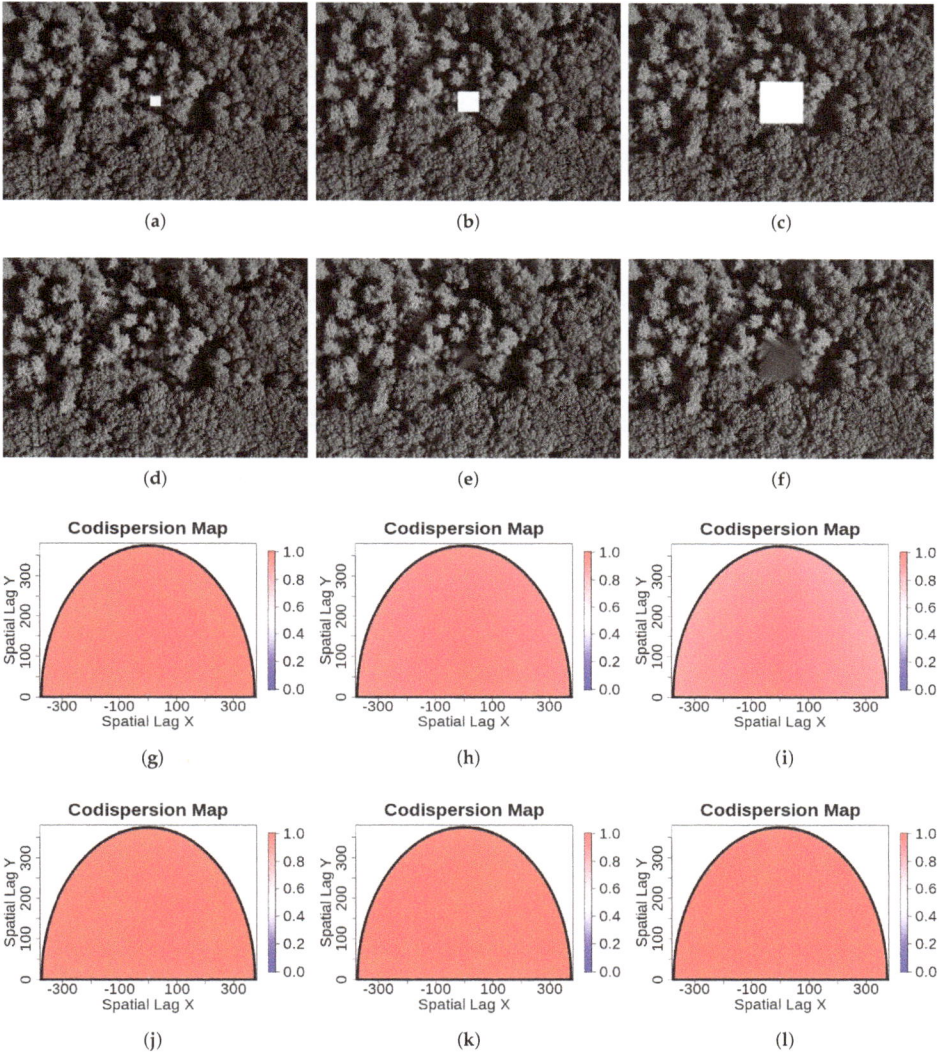

Figure 10. Contamination of the reference Figure 1a by gaps resulting from clusters of missing observations. Images (**a–c**) contain only one missing block in the center of the image of sizes 200 × 200, 400 × 400, y 800 × 800, respectively. These missing data were filled in images (**d–f**) using the imputation algorithm described in the Appendix. Images (**g–l**) are the corresponding codispersion maps between Figure 1a and the imputed images (**d–f**).

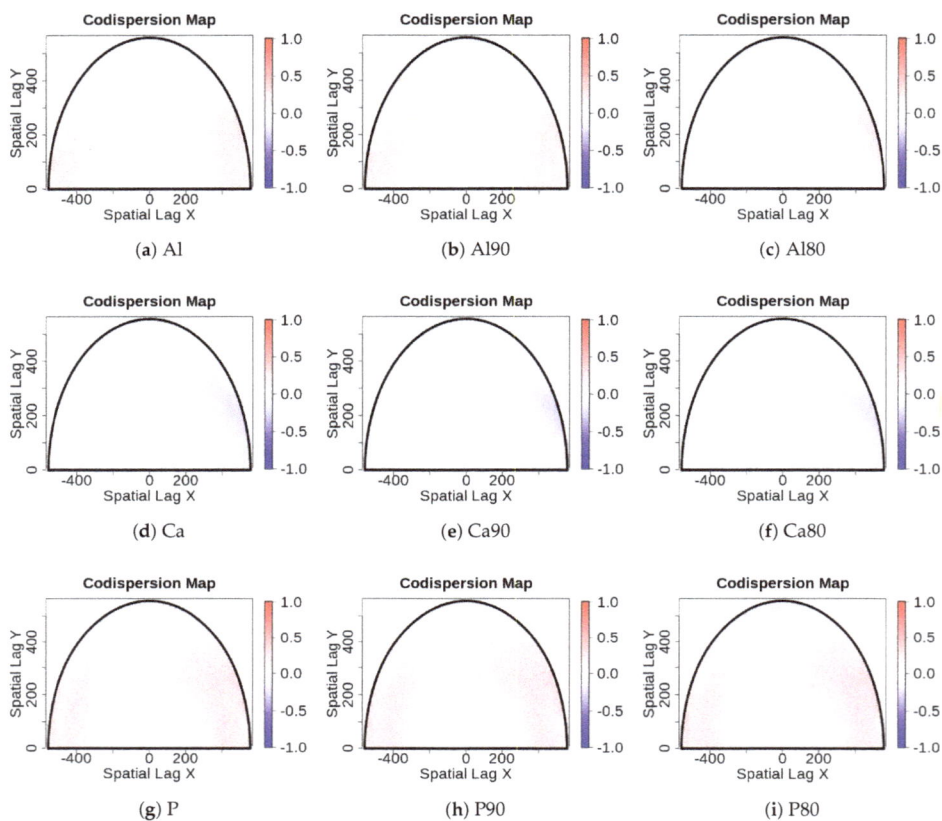

Figure 11. Codispersion between species *A. blackiana* and soil chemistry variables; (Al (**a–c**); Ca (**d–f**) and P (**g–i**)). Soil data were unthinned (**a,d,g**); thinned 10% (**b,e,h**); or thinned 20% (**c,f,i**).

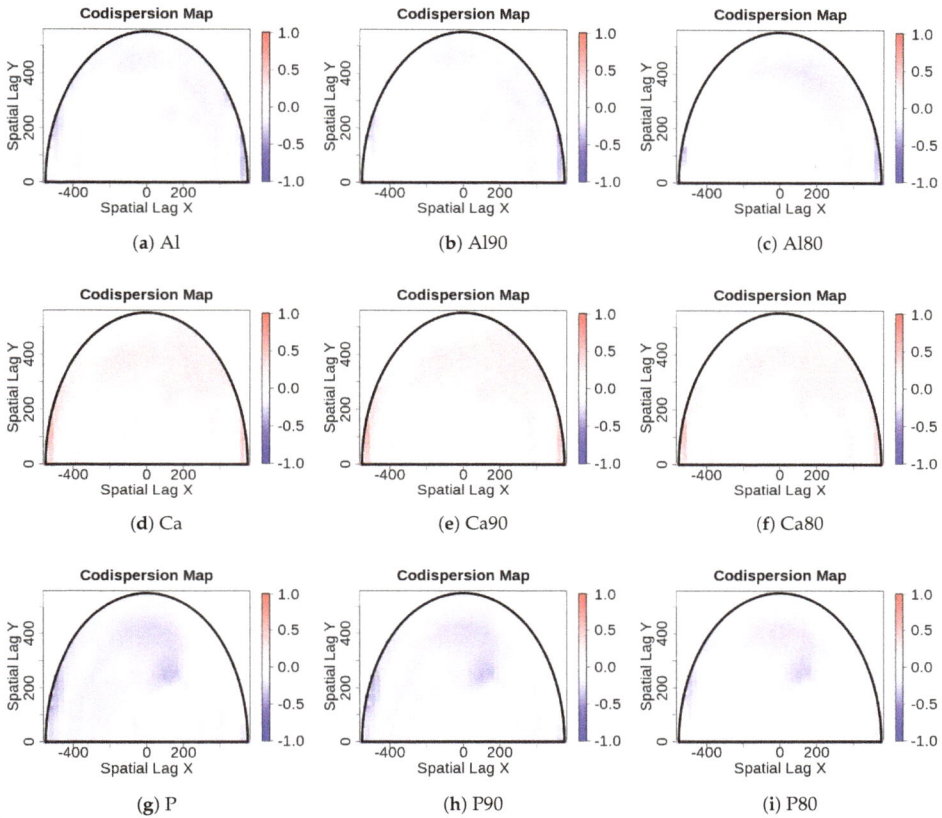

Figure 12. Codispersion between species *P. armata* and soil chemistry variables; (Al (**a–c**); Ca (**d–f**) and P (**g–i**)). Soils data were unthinned (**a,d,g**), thinned 10% (**b,e,h**), or thinned 20% (**c,f,i**).

4. Discussion

The methods and examples developed in this paper improve our understanding of the behavior of the codispersion coefficient when data are noisy or have been contaminated by various types of errors common in remotely-sensed images or in interpolated and predicted (e.g., kriged) surfaces. The codispersion coefficient appears to be robust for small percentages of contamination (<15%) but always leads to an underestimation of the codispersion between the datasets. As the percentage of contamination increases, codispersion decreases uniformly for all directions on the plane, thus the types of noise considered in this paper did not affect the codispersion in any particular direction(s). Although the performance of codispersion for directional noise was explored in [3,25], directional noise has not yet been observed in real datasets.

When applied to data collected from large forest plots, codispersion has been shown to be useful for describing scales of covariation in two or more variables across complex spatial gradients (e.g., [4,5]). Our ability to detect such spatial pattern depends on the grain of spatial variation in the data and how this compares to the lag sizes used in the codispersion analysis. For example, the complete loss of correlation between the two images in Figure 2 with only a small degree of contamination highlights the importance of considering the spatial grain of the datasets relative to that of the noise-inducing processes. The coarser-grained spatial pattern in the forest images is retained, even under contamination, whereas the spatial dependence in the images in Figure 2 is at a smaller grain than the extent of the image, which is relatively heavily disturbed by the salt-and-pepper noise.

The imputation algorithm described in Appendix A seems to be a promising technique to handle blocks of missing observations. Several aspects of it are worth exploring with future research. These include the success of the algorithm in recovering missing observations as a function of the block size; how to select the number of neighbors to be considered in the AR-2D process; and the similarity between the texture of the imputed observations and the texture of the reference image. For simplicity and without loss of generality, the missing blocks we illustrated were square regions located in the center of the image, but certainly Algorithm 2 could be extended to other types of regions located anywhere in the image.

More general aspects of codispersion analysis are in need of further exploration and testing. First, it will be of interest to study the results of codispersion analysis of rasterized images. This is because rasterization of images is widespread and common rasterization methods rarely, if ever, preserve the original spatial correlation of each process. The development of a new rasterization method that preserves better the spatial correlation within processes could follow [26]. Second, the computation of codispersion maps is computationally expensive. Thus, the development of efficient algorithms capable of creating codispersion maps for large images is still needed.

5. Conclusions

The codispersion map is a useful tool that illustrates those directions for which the codispersion coefficient between two spatial processes attains its maximum and minimum values. When the direction of interest in unknown, the codispersion map also visually and concisely summarizes the correlation between two processes in a plane. When data are noisy or have some degree of observation or process error our results suggest that:

(1) The codispersion coefficient is robust to small percentages of contamination (less than 15%).
(2) The codispersion coefficient decreases as the percentage of contamination increases no matter the type of noise or direction.
(3) For data collected from large forest plots, the codispersion coefficient and the associated codispersion map provide useful information to describe covariation in the data across complex spatial gradients or patterns.
(4) An imputation algorithm can be used to smoothly fill blocks of missing observations with little impact on the codispersion coefficient.

The development of codispersion maps for large data sets can be addressed by using the effective sample size for spatial variables, recently proposed in [27].

Supplementary Materials: Analyses were done using the R software system, version 3.3.1 [28]. The images and all the code used in this paper are available from https://github.com/JAcostaS/Code-and-Example-Codismap.git. Barro Colorado Island (BCI) vegetation and soils data are available from http://ctfs.si.edu/webatlas/datasets/bci/.

Author Contributions: R.V. developed the theoretical formalism of the codispersion map, J.A. worked on the computational aspects related to the construction of the codispersion map and the imputation algorithm. H.B. and B.C. performed the numeric calculations and the display of the example related with individual forest trees and soil nutrient concentrations. A.M.E. conceived this research, plan the structure of the paper and selected the appropriate real examples. All authors wrote parts of the paper, provided critical feedback and helped shape the research, analysis and manuscript.

Funding: This research was funded by AC3E, grant number FB-0008, Chile.

Acknowledgments: R.V. was partially supported by AC3E, FB-0008, Valparaíso, Chile. J.A. was supported by Pontificia Universidad Católica de Valparaíso, Grant 039.320/2018. A.M.E.'s participation in this project was supported by Harvard University and the Universidad Técnica Federico Santa María, and Hannah Buckley. and Bradley Case's work on this project in Chile also was supported by the Universidad Técnica Federico Santa María. Vegetation data from BCI are part of the BCI forest dynamics research project founded by S. P. Hubbell and R. B. Foster and now managed by R. Condit, S. Lao, and R. Perez through the Center for Tropical Forest Science (CTFS) and the Smithsonian Tropical Research Institute (STRI) in Panamá. Numerous organizations have provided funding to support this long-term study, principally the US National Science Foundation (NSF), and hundreds of field workers have contributed to mapping, measuring and monitoring the vegetation. Jim Dalling, Robert John, Kyle Harms, Robert Stallard, Joe Yavitt, Paolo Segre, and Juan Di Trani sampled the soils at BCI. Collection and initial analysis of the BCI data were supported by NSF grants 021104, 021115, 0212284, 0212818 and 0314581, the STRI Soils Initiative, and CTFS. This paper is a publication of the Harvard Forest Long-Term Ecological Research Site, supported by the US National Science Foundation.

Conflicts of Interest: The authors declare no conflict of interest.

Appendix A. Image Imputation Algorithm

The algorithm described below is based on the fact that it is possible to represent any image by using unilateral AR-2D processes [17,29]. The generated image is called a local AR-2D approximated image by using blocks.

Let $Z = \{Z_{r,s} : 0 \leq r \leq M-1, 0 \leq s \leq N-1\}$ be an original image, and let X the original image corrected by the mean. That is, $X_{r,s} = Z_{r,s} - \overline{Z}$, for all $0 \leq r \leq M-1, 0 \leq s \leq N-1$, and for which $\overline{Z}$ is the mean of Z.

Following [30], assume that X follows a causal AR-2D process of the form

$$X_{r,s} = \phi_1 X_{r-1,s} + \phi_2 X_{r,s-1} + \phi_3 X_{r-1,s-1} + \varepsilon_{r,s},$$

where $(r,s) \in \mathbb{Z}^2$, $(\varepsilon_{r,s})_{(r,s)\in\mathbb{Z}^2}$ is Gaussian white noise, and ϕ_1, ϕ_2, and ϕ_3 are the autoregressive parameters.

Let $4 \leq k \leq \min(M,N)$. For simplicity, we consider that the images to be processed are arranged in such a way that the number of columns minus one and the number of rows minus one are multiples of $k-1$; Then, we define the $(k-1) \times (k-1)$ block (i_b, j_b) of the image X by

$$B_X(i_b, j_b) = \{X_{r,s} : (k-1)(i_b-1)+1 \leq r \leq (k-1)i_b, (k-1)(j_b-1)+1 \leq s \leq (k-1)j_b\},$$

for all $i_b = 1, \cdots, [(M-1)/(k-1)]$ and for all $j_b = 1, \cdots, [(N-1)/(k-1)]$, where $[\cdot]$ denotes the integer part. The $M' \times N'$ approximated image $\widehat{Z}$, where $M' = [(M-1)/(k-1)](k-1)+1$ and $N' = [(N-1)/(k-1)](k-1)+1$ can be obtained by the following algorithm.

Algorithm 1 Approximated AR-2D Image.

Input: An original image Z of size $M \times N$.

Output: An approximated $\widehat{Z}$ of size $M' \times N'$.

1: **for each block** $B_X(i_b, j_b)$ **do**
2:     Compute the least square (LS) estimators of ϕ_1, ϕ_2 and ϕ_3 associated with block $B_X(i_b, j_b)$.
3:     Define $\widehat{X}$ on the block $B_X(i_b, j_b)$ by

$$\widehat{X}_{r,s} = \widehat{\phi}_1(i_b, j_b) X_{r-1,s} + \widehat{\phi}_2(i_b, j_b) X_{r,s-1} + \widehat{\phi}_3(i_b, j_b) X_{r-1,s-1},$$

    where $(k-1)(i_b-1)+1 \leq r \leq (k-1)i_b, (k-1)(j_b-1)+1 \leq s \leq (k-1)j_b$, and $\widehat{\phi}_1(i_b, j_b)$, $\widehat{\phi}_2(i_b, j_b)$, and $\widehat{\phi}_3(i_b, j_b)$ are the LS estimators of ϕ_1, ϕ_2 and ϕ_3 respectively.
4: **end for**
5:     The approximated image $\widehat{Z}$ of Z is:

$$\widehat{Z}_{r,s} = \widehat{X}_{r,s} + \overline{Z}, \quad 0 \leq r \leq M'-1, 0 \leq s \leq N'-1.$$

6: **Return** $\widehat{Z}$.

Now suppose that image Z has a rectangular block of missing values. Without loss of generality, assume that the rectangular block of missing values is of size $(K-1) \times (K-1)$. Furthermore, in each border, $X^{(l)}$, $l = 1, 2, 3, 4$, is defined as a block of information of Z of size $K \times K$, such as appears in Figure A1.

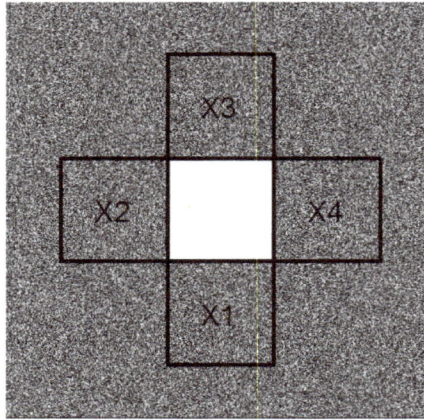

Figure A1. Block of missing values.

In addition, assume that all l, $X^{(l)}$ are represented by a AR-2D model of the form

$$X_{r,s}^{(l)} = \phi_1^{(l)} X_{r-1,s}^{(l)} + \phi_2^{(l)} X_{r,s-1}^{(l)} + \phi_3^{(l)} X_{r-1,s-1}^{(l)} + \varepsilon_{r,s}^{(l)}, \quad l = 1, 2, 3, 4,$$

where $\phi_1^{(l)}$, $\phi_2^{(l)}$, and $\phi_3^{(l)}$ are estimated using the block $X^{(l)}$ for $l = 1, 2, 3, 4$, respectively. Then, the prediction model is

$$\widehat{X}_{r+i,s+j}^{(l)} = \begin{cases} \widehat{\phi}_1^{(l)} \widehat{X}_{r+i-1,s+j}^{(l)} + \widehat{\phi}_2^{(l)} \widehat{X}_{r+i,s+j-1}^{(l)} + \widehat{\phi}_3^{(l)} \widehat{X}_{r+i-1,s+j-1}^{(l)} & ; \quad (r+i,s+j) \notin A^{(l)}, \\ X_{r+i,s+j}^{(l)} & ; \quad (r+i,s+j) \in A^{(l)}, \end{cases}$$

where $A^{(l)}$ is the index set for which $X^{(l)}$ is known and $i, j = 1, \ldots, K$. The prediction algorithm is the following

---

Algorithm 2 Prediction Algorithm.

Input: An image Z with a missing block, and K.

Output: Image Z without missing values.

1: Get a sub-image X of Z of size $3K \times 3K$, so that the missing data is in the center of X.

2: Get $X^{(l)}$, for $l = 1, 2, 3, 4$, and reverse the order of the rows in $X^{(3)}$ and the columns in $X^{(4)}$, i.e.,

$$X_{i,j}^{(3)} = X_{K+1-i,j}^{(3)}, \text{ and } X_{i,j}^{(4)} = X_{i,K+1-j}^{(4)}$$

3: Compute $\widehat{\phi}_1^{(l)}$, $\widehat{\phi}_2^{(l)}$ and $\widehat{\phi}_3^{(l)}$ for $l = 1, 2, 3, 4$.

4: Let $K_2 = K$.

5: **while** $K_2 > 0$. **do**

6:     **for** $j = 1$ **until** $j = K - 1$ **do**

7:         Compute:

$$
\begin{aligned}
X_{K+1,K+j} &= \widehat{\phi}_1^{(1)} X_{K,K+j} + \widehat{\phi}_2^{(1)} X_{K+1,K+j-1} + \widehat{\phi}_3^{(1)} X_{K,K+j-1} \\
X_{K+j,K+1} &= \widehat{\phi}_1^{(2)} X_{K+j-1,K+1} + \widehat{\phi}_2^{(2)} X_{K+j,K} + \widehat{\phi}_3^{(2)} X_{K+j-1,K} \\
X_{2K-1,K+j} &= \widehat{\phi}_1^{(3)} X_{2K,K+j} + \widehat{\phi}_2^{(3)} X_{2K-1,K+j-1} + \widehat{\phi}_3^{(3)} X_{2K,K+j-1} \\
X_{K+j,2K-1} &= \widehat{\phi}_1^{(4)} X_{K+j-1,2K-1} + \widehat{\phi}_2^{(4)} X_{K+j,2K} + \widehat{\phi}_3^{(4)} X_{K+j-1,2K}
\end{aligned}
$$

▷ For those points that the estimation is repeated consider the average of both estimations. These points are obtained for $j = 1$ and $j = K - 1$.

    end for

8:     Put $K_2 = K_2 - 2$ and $K = K + 1$.

end while

9: Replace the NA values of Z by X.

10: **Return** Z.

References

1. Fortin, M.J.; Dale, M. *Spatial Analysis: A Guide for Ecologists*; Cambridge University Press: Cambridge, UK, 2005; pp. 5–11.

2. Ellison, A.M.; Gotelli, N.J.; Hsiang, N.; Lavine, M.; Maidman, A. Kernel density estimation of 2-dimensional spatial Poisson point processes from *k*-tree sampling. *J. Agric. Biol. Environ. Stat.* **2014**, *19*, 357–372. [CrossRef]

3. Vallejos, R.; Osorio, F.; Mancilla, D. The codispersion map: A graphical tool to visualize the association between two spatial processes. *Stat. Neerl.* **2015**, *69*, 298–314. [CrossRef]

4. Buckley, H.L.; Case, B.S.; Ellison, A.M. Using codispersion analysis to characterize spatial patterns in species co-occurrences. *Ecology* **2016**, *97*, 32–39. [CrossRef] [PubMed]

5. Buckley, H.L.; Case, B.S.; Zimmermann, J.; Thompson, J.; Myers, J.A.; Ellison, A.M. Using codispersion analysis to quantify and understand spatial patterns in species-environment relationships. *New Phytol.* **2016**, *211*, 735–749. [CrossRef] [PubMed]

6. Case, B.S.; Buckley, H.L.; Barker Plotkin, A.; Ellison, A.M. Using codispersion analysis to quantify temporal changes in the spatial pattern of forest stand structure. *Chil. J. Stat.* **2016**, *7*, 3–15.

7. Ellison, A.M.; Osterweil, L.J.; Hadley, J.L.; Wise, A.; Boose, E.; Clarke, L.; Foster, D.R.; HAnson, A.; Jensen, D.; Kuzeja, P.; et al. Analytic webs support the synthesis of ecological datasets. *Ecology* **2006**, *87*, 1345–1358. [CrossRef]

8. Matheron, G. *Les Variables Régionalisées et leur Estimation*; Masson: Paris, France, 1965.

9.   Ojeda, S.; Vallejos, R.; Lamberti, P. Measure of similarity between images based on the codispersion coefficient. *J. Electron. Imaging* **2012**, *21*, 023019. [CrossRef]

10.   Anselin, L. Local indicators of spatial association–LISA. *Geogr. Anal.* **1995**, *27*, 93–115. [CrossRef]

11.   Fox, A.J. Outliers in time series. *J. R. Stat. Soc. B* **1972**, *34*, 350–363.

12.   Huang, S.; Zhu, J. Removal of salt-and-pepper noise based on compressed sensing. *Electron. Lett.* **2010**, *46*, 1198–1199. [CrossRef]

13.   McQuarrie, A.D.; Tsai, C. Outlier detections in autoregressive models. *J. Comput. Graph. Stat.* **2003**, *12*, 450–471. [CrossRef]

14.   Gneiting, T.; Kleiber, W.; Schlather, M. Matérn cross-covariance functions for multivariate random fields. *J. Am. Stat. Assoc.* **2010**, *105*, 1167–1177. [CrossRef]

15.   Schlather, M.; Malinowski, A.; Oesting, M.; Boecker, D.; Strokorb, K.; Engelke, S.; Martini, J.; Ballani, F.; Moreva, O.; Auel, J.; et al. RandomFields: Simulation and Analysis of Random Fields. R Package Version 3.1.50. 2017. Available online: https://cran.r-project.org/package=RandomFields (accessed on 1 April 2018).

16.   Allende, H.; Galbiati, J.; Vallejos, R. Robust image modeling on image processing. *Pattern Recognit. Lett.* **2001**, *22*, 1219–1231. [CrossRef]

17.   Ojeda, S.; Vallejos, R.; Bustos, O. A new image segmentation algorithm with applications to image inpainting. *Comput. Stat. Data Anal.* **2010**, *54*, 2082–2093. [CrossRef]

18.   Minasny, B.; McBratney, A.B. The Matérn function as a general model for soil variograms. *Geoderma* **2005**, *128*, 192–207. [CrossRef]

19.   Condit, R. *Tropical Forest Census Plots*; Springer: Berlin, Germany, 1998.

20.   Hubbell, S.P.; Condit, R.; Foster, R.B. Barro Colorado Forest Census Plot Data. 2005. Available online: http://ctfs.si.edu/webatlas/datasets/bci (accessed on 18 February 2018).

21.   Hubbell, S.P.; Foster, R.B.; O'Brien, S.T.; Harms, K.E.; Condit, R.; Wechsler, B.; Wright, S.J.; Loo de Lao, S. Light gap disturbances, recruitment limitation, and tree diversity in a neotropical forest. *Science* **1998**, *283*, 554–557. [CrossRef]

22.   John, R.; Dalling, J.W.; Harms, K.E.; Yavitt, J.B.; Stallard, R.F.; Mirabello, M.; Hubbell, S.P.; Valencia, R.; Navarrete, H.; Vallejo, M.; et al. Soil nutrients influence spatial distributions of tropical trees. *Proc. Natl. Acad. Sci. USA* **2007**, *104*, 864–869. [CrossRef]

23.   Ribeiro, P.J., Jr.; Diggle, P.J. geoR: A package for geostatistical analysis. *R-News* **2001**, *1*, 15–18.

24.   Wang, J.-F.; Zhang, T.-L.; Fu, B.-J. A measure of spatial stratified heterogeneity. *Ecol. Indic.* **2016**, *67*, 250–256. [CrossRef]

25.   Vallejos, R.; Mancilla, D.; Acosta, J. Image similarity assessment based on measures of spatial association. *J. Math. Imaging Vis.* **2016**, *56*, 77–98. [CrossRef]

26.   Goovaerts, P. Combining Areal and Point Data in Geostatistical Interpolation: Applications to Soil Science and Medical Geography. *Math. Geosci.* **2010**, *42*, 535–554. [CrossRef] [PubMed]

27.   Acosta, J.; Vallejos, R. Effective sample size for spatial regression processes. *Electron. J. Stat.* **2018**, *12*, 3147–3180. [CrossRef]

28.   R Development Core Team. *R: A Language and Environment for Statistical Computing*; R Foundation for Statistical Computing: Vienna, Austria, 2016. Available online: http://www.R-project.org (accessed on 22 March 2018).

29.   Ver Hoef, J.M.; Peterson, E.E.; Hooten, M.B.; Hanks, E.M.; Fortin, M.J. Spatial autoregressive models for statistical inference from ecological data. *Ecol. Monogr.* **2018**, *88*, 36–59. [CrossRef]

30.   Bustos, O.; Ojeda, S.; Vallejos, R. Spatial ARMA models and its applications to image filtering. *Braz. J. Prob. Stat.* **2009**, *23*, 141–165. [CrossRef]

forests

Article

Estimating Individual Tree Height and Diameter at Breast Height (DBH) from Terrestrial Laser Scanning (TLS) Data at Plot Level

Guangjie Liu [1,2,3], Jinliang Wang [1,2,3,*], Pinliang Dong [4], Yun Chen [1,2,3] and Zhiyuan Liu [1,2,3]

[1] College of Tourism and Geographic Sciences, Yunnan Normal University, Kunming 650500, China;
     guangjiers@gmail.com (G.L.); cqchenyun@126.com (Y.C.); liuzhiyuangis@163.com (Z.L.)
[2] Key Laboratory of Resources and Environmental Remote Sensing for Universities in Yunnan,
     Kunming 650500, China
[3] Center for Geospatial Information Engineering and Technology of Yunnan Province, Kunming 650500, China
[4] Department of Geography and the Environment, University of North Texas, 1155 Union Circle #305279,
     Denton, TX 76203, USA; pdong@unt.edu
*   Correspondence: wang_jinliang@hotmail.com; Tel.: +86-0871-6594-1202

Received: 12 June 2018; Accepted: 2 July 2018; Published: 4 July 2018

Abstract: Abundant and refined structural information under forest canopy can be obtained by using terrestrial laser scanning (TLS) technology. This study explores the methods of using TLS to obtain point cloud data and estimate individual tree height and diameter at breast height (DBH) at plot level in regions with complex terrain. Octree segmentation, connected component labeling and random Hough transform (RHT) are comprehensively used to identify trunks and extract DBH of trees in sample plots, and tree height is extracted based on the growth direction of the trees. The results show that the topography, undergrowth shrubs, and forest density influence the scanning range of the plots and the accuracy of feature extraction. There are differences in the accuracy of the results for different morphological forest species. The extraction accuracy of Yunnan pine forest is the highest (DBH: Root Mean Square Error (RMSE) = 1.17 cm, Tree Height: RMSE = 0.54 m), and that of *Quercus semecarpifolia* Sm. forest is the lowest (DBH: RMSE = 1.22 cm, Tree Height: RMSE = 1.23 m). At plot scale, with the increase of the mean DBH or tree height in plots, the estimation errors show slight increases, and both DBH and height tend to be underestimated.

Keywords: diameter at breast height (DBH); tree height; random Hough transform; point cloud; terrestrial laser scanning

1. Introduction

Earth's forests, which cover 30% of the total land area, are dynamic systems that are constantly in a state of change and drive/respond to the changes taking place in our environment. Tree height, diameter at breast height (DBH) and other forest structure parameters are examples of the important basic data recovered from a traditional forest resource survey. They are of great significance for the research on forest biomass estimation, forest carbon cycle, carbon flow and global climate change. With the development of remote sensing technology, especially the technology of Light Detection and Ranging (LiDAR), many research results have been obtained by using remotely sensed data to extract information on forest structure parameters. There are mainly two ways for estimating parameters of forest structure using traditional passive optical imaging: (1) the correlation between forest structural parameters and spectral information is established by using multi-spectral characteristics of optical remote sensing data [1,2]; and (2) forest structure parameters are extracted using high spatial resolution image texture features [3–5]. However, due to the complicated structure of forest canopy cover, atmospheric scattering, and topography, it is difficult for optical remote sensing to provide accurate information about the vertical distribution of the forests [6,7]. As microwaves can penetrate dense canopy to obtain information on the branches and trunks below forest canopies, synthetic aperture radar (SAR) is more advantageous than passive optical remote sensing methods to detect forest structure parameters and biomass [4]. Backscattering mechanism of SAR data [8–10] and interferometric synthetic aperture radar (InSAR) [11,12], polarimetric synthetic aperture radar interferometry (POLinSAR) [13,14] and polarization coherence tomography (PCT) [15,16] techniques all have obtained many research results on forest structural parameters. Meanwhile, LiDAR has been intensively applied to the study of forest structural parameters. As space-borne LiDAR can obtain a wide extent of tree height information, it has been applied to studies on large-scale forest biomass [17,18] and forest canopy height [19,20]. However, the new generation of LiDAR satellite ICESat-2 has not yet been launched, and the lack of spaceborne LiDAR data remains a limiting factor [21]. Airborne LiDAR has the ability to obtain the vertical structure of large areas of forest, but it usually cannot reflect detailed structural information under tree canopy [22]. Compared with the above two LiDAR platforms, terrestrial laser scanning (TLS) obtains high density point clouds and can get more detailed information on forest internal structure, including tree location, DBH, tree height, crown width, and other biophysical parameters.

TLS is a laser-based instrument that measures its surroundings using LiDAR for range measurement and precise angular measurements through the optical beam deflection mechanism to derive 3D point observations from the object surfaces [23]. The high-density point cloud data obtained by TLS is widely used and researched in many fields such as engineering surveys [24,25], Earth sciences [26,27], natural disasters [28–30], coastline erosion [31–33], vegetation monitoring [34,35], and digital terrain mapping [35,36]. In recent years, TLS has been increasingly applied to forest resource surveys, forest management and planning [37,38]. Among a variety of forest structural parameters, DBH and tree height are the most important ones obtained in forest resource surveys. They can provide not only structural parameters of individual trees but also information and data on sample plot level, which are of great significance for the study of forest carbon storage and biomass estimation. Many researchers have conducted investigations on how to extract DBH, tree height and others structural parameters using TLS data efficiently and accurately.

In terms of methodology, the methods for automatically extracting DBH from TLS data mainly include Hough transform [22,39,40], circle fitting algorithm [41–47], and cylinder fitting algorithm [42,48,49]. Li [40] used the Hough transform method to detect circles on rasterized point cloud data to estimate DBH and tree height. Liu et al. [22] applied the Hough transform method to natural forest and plantation in Puer City, China, and concluded that TLS data could be used to extract DBH (RMSE = 2.18 cm, R^2 = 0.91). Bienert et al. [41] used a method for fitting circles to extract DBH of trees accurately, and concluded that the tree trunks blocked each other when the tree density was high, which resulted in the reduction of DBH extraction precision or even led to the unrecognizable trees. Moskal et al. [38] used the method of cylindrical fitting to extract DBH, with an RMSE of 9.17 cm. The main reasons for the relatively low accuracy were the poor visibility of the scanning station and the blockage of individual tree trunks. Due to mutual occlusion between the canopy of individual trees, tree heights extracted from TLS point cloud data are always lower than the measured values [22]. The most commonly used method for tree height extraction is to obtain the highest point over the ground within a certain range of a single tree, and use the height of the highest cloud point as the tree height [50]. In order to improve the extraction precision of tree height, most studies have employed the circle fitting method to determine the growth direction of the tree trunk, and calculate the tree height along the growth direction of the tree trunk [22,40,51]. In order to improve the efficiency of the algorithm, the method of extracting DBH based on circle detection or circle fitting needs to rasterize the point cloud data, which reduces the availability of data and the extraction accuracy [22].

As far as study areas are concerned, most of the studies on the extraction of forest structural parameters from TLS data focus on plantations of single forest types or a small amount of natural forests, and research on tree height and DBH extraction of natural forests from typical tree species in a particular area is lacking. Also, most of studies on DBH and tree height inversion have been carried out at scales of individual woods, and studies at scales of forest sample plots with multiple tree species and multi-aged forests are still lacking.

To improve the efficiency and accuracy of forest resource surveys, this study explores methods for extracting tree height and DBH at plot level in complex terrain and different sub-wooded environments using TLS data. Four types of dominant forest species (*Pinus yunnanensis* Franch., *Pinus densata* Mast., *Picea* Mill. & *Abies fabri* (Mast.) Craib, *Quercus semecarpifolia* Sm.) are investigated in Shangri-La, northwest of Yunnan, China. Identification of individual trees and extraction of DBH from TLS point cloud data are implemented by using octree segmentation, connected component labeling (CCL) and Random Hough Transform (RHT), following the tree growth direction obtained from TLS point data. Based on the extracted individual tree DBH and tree height, the average DBH and the average tree height are obtained by method of square average.

2. Materials and Methods

2.1. Study Area and Sample Plots

Shangri-La is located in the northwestern part of Yunnan Province, China, the eastern part of Diqing Tibetan Autonomous Prefecture, between $26°52'$~$28°52'$ N and $99°22'$~$100°19'$ E with an area of 11,613 km^2. It is one of the largest county-level administrative areas in Yunnan Province (Figure 1). With elevations over 3000 m above sea level in most areas of Shangri-La, the main landform types in the region are subalpine and alpine, which determines the distribution of cold-temperate coniferous forests and temperate-cool coniferous forests in the area. The area of woodland is 962,159.3 hectares in Shangri-La, and total volume of living wood is 133,224,410 m^3; the forest coverage rate is 76.00%, and the forest greening rate is 83.19%. *Quercus semecarpifolia*, *Pinus yunnanensis*, *Pinus densata*, and *Picea* & *Abies fabri* (including *Abies georgei* Orr, *Abies delavayi* Franch., and *Picea likiangensis* (Franch.) E.Pritz.) account for 90.8% of the total area of arbors in Shangri-La.

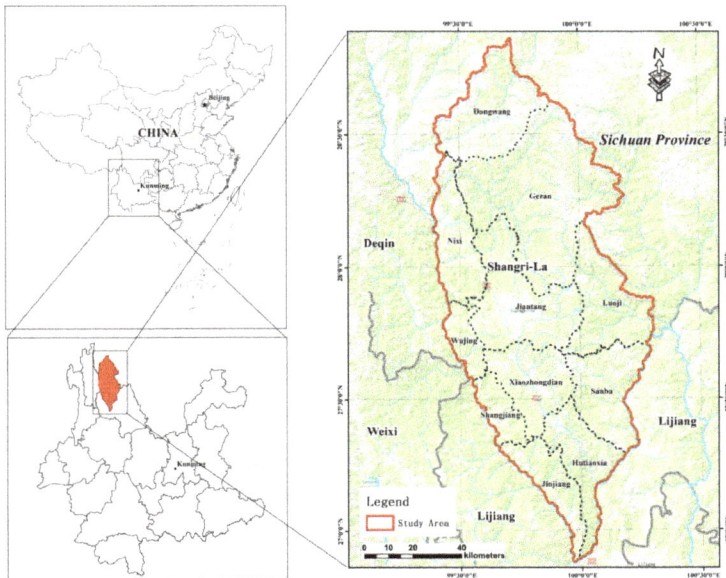

Figure 1. Study Area.

This study used the Leica P40 to acquire high-precision 3D point cloud data. The P40 is Leica's latest generation 3D laser scanning device for fast, high-density point cloud and panoramic image collection. The main performance indicators of the device are shown in Table 1.

Table 1. Main performance indicators of Leica P40.

Indicators	Descriptions
Range Accuracy	1.2 mm + 10 ppm
3D position Accuracy	3 mm @ 50 m 6 mm @ 100 m
Wavelength	1550nm (invisible); 658 nm (visible)
Scan Rate	Up to 1,000,000 points per second
Field-of-View	360° (Horizontal); 290° (Vertical)
Range and Reflectivity	Minimum range: 0.4 m Maximum range at reflectivity: 120 m (8%), 180 m (18%), 270 m (34%)
Range Noise	0.4 mm RMS at 10 m 0.5 mm RMS at 50 m

Point cloud data were obtained at three different times (August 2016, July 2017 and September 2017) respectively. Because the point cloud data of all trees in a sample plot cannot be acquired by only one scanning station, the method of measuring from multiple stations is used in the study. In each plot, five stations (four stations in some samples) were scanned. One station was in the center of the plot with refined scanning method for 10-min scanning, and panoramic photos were obtained at the same time. Other stations were set up on the edge of the sample plot with a 5-min scanning. The study obtained 196 stations of LiDAR point cloud data in 39 forest sample plots (Table 2), which were distributed in various townships in Shangri-La (Figure 2).

In order to obtain a sufficient amount of data for verification and ensure the reliability of research results, we used DBH rulers, Trueyard SP1500H laser rangefinder and steel tape to obtain forest structural parameters in all 39 forest sample plots (Figure 2). The range of forest plots varies according to topography and forest density, but the diameter of each plot is not less than 40 m. With the topographical conditions permitting, the range of the sample plot was expanded as much as possible to obtain more data of the tree and to verify the range and accuracy of laser scanning.

Table 2. Number of different types of forest sample plots in the study.

Dominant Forest Species	Age of Stand	Number of Sample Plots	Number of Stations	Average Altitude (Unit: m)	Average Slope (Unit: Degree)
Quercus semecarpifolia Sm.	Young	1	5	3892	11.0
	Middle	2	10	3673	15.0
	Mature	1	4	3723	30.0
Pinus densata Mast.	Young	3	17	3225	16.0
	Middle	4	20	3210	16.4
	Mature	2	9	3128	23.5
Pinus yunnanensis Franch.	Young	3	14	2538	19.3
	Middle	5	25	2692	15.3
	Mature	8	43	2316	13.9
Picea Mill. & *Abies fabri* (Mast.) Craib	Young	2	10	3453	23.3
	Middle	4	20	3604	13.0
	Mature	4	19	3680	15.6

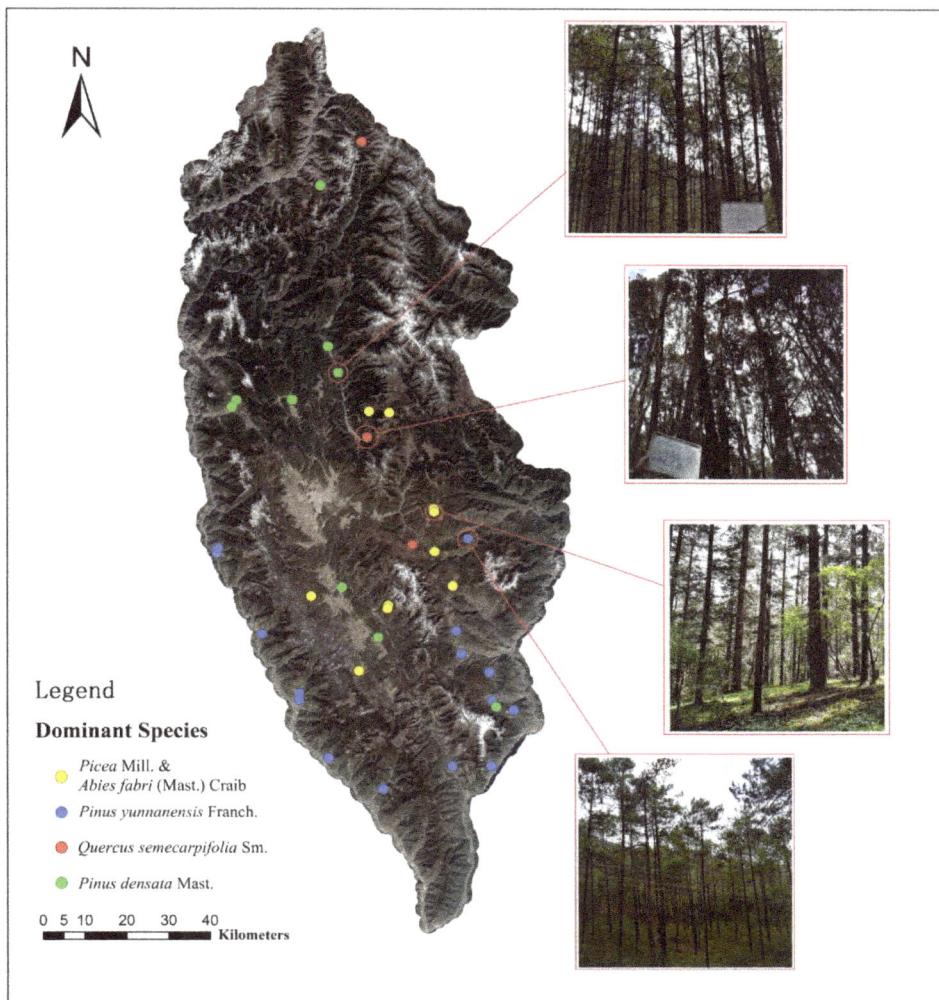

Figure 2. Forest sample plots in Shangri-La, Yunnan, China.

2.2. Data Acquisition and Processing

The main research process includes point cloud data preprocessing, normalization of point cloud height, point cloud segmentation, trunk identification, and tree height and DBH extraction. A flowchart detailing the methods in this study is shown in Figure 3. First, a software, Leica Cyclone, is used to stitch multi-site point cloud data based on the Leica 4.5″ circular black & white target. Because there is a lot of redundancy in multi-site point cloud data, the software also is used to deduct data so that we can reduce the time cost in data processing under the premise of ensuring data extraction accuracy.

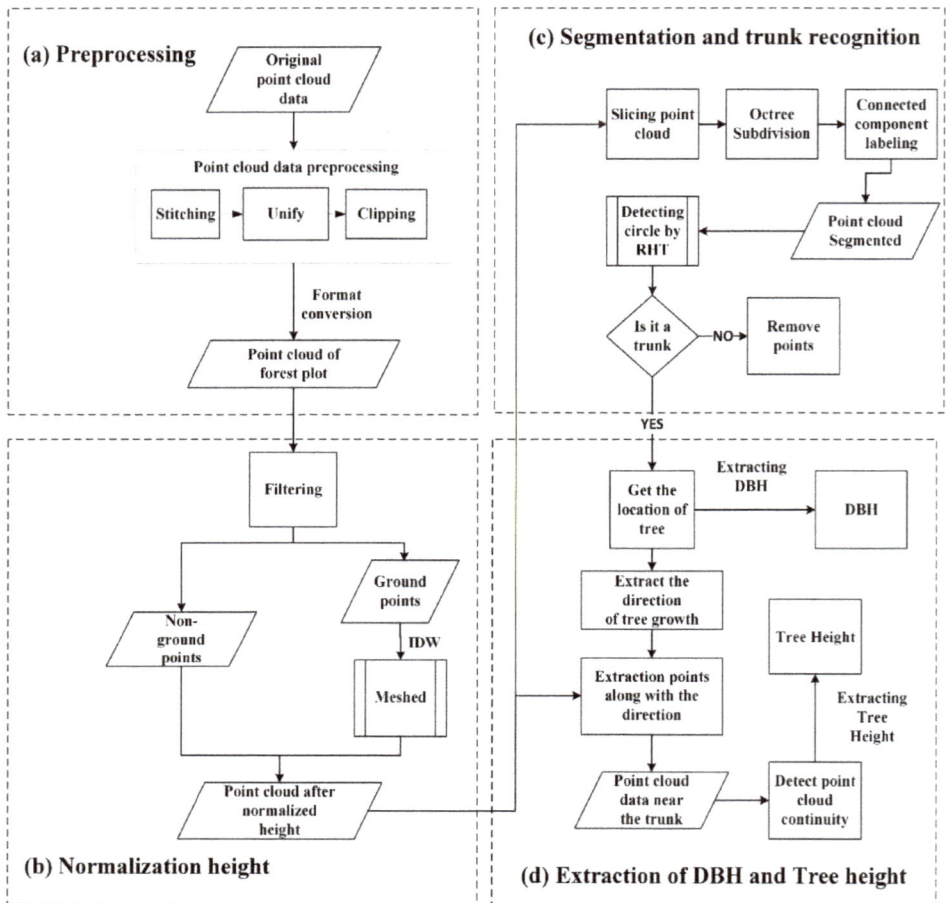

Figure 3. Flowchart detailing the methods in this study. (**a**) In order to remove useless data and reduce the amount of data, point cloud data needs to be preprocessed; (**b**) Normalization of points height facilitates the extraction of DBH and tree height; (**c**) The slicing and segmenting point clouds can improve the efficiency and accuracy of trunk recognition; (**d**) According to the trunk position, directly we extract or fit the DBH. Tree heights are obtained based on the tree growth direction and continuity detecting.

2.2.1. Normalization of Point Cloud Height

A morphological filtering method [52] is used to separate ground points from non-ground points. The main idea of morphological filtering is to use the corrosion and expansion operations in mathematical morphology to remove the higher point cloud in the point cloud and keep the lower point cloud to achieve the purpose of extracting ground points [53]. Ground points are interpolated and meshed by Inverse Distance Weighting (IDW) method. Finally, using the generated grid of ground, points heights are normalized to eliminate the difference in tree height caused by differences in elevation (Figure 4).

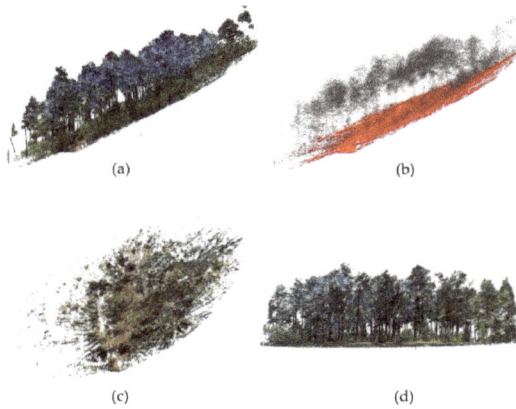

Figure 4. Normalization of point cloud height. (**a**) Original point cloud data acquired using TLS; (**b**) Filtering results with ground points in red and non-ground points in gray; (**c**) Ground points with RGB color; (**d**) Points with normalized height.

2.2.2. Slicing Point Clouds

When obtaining 3D point cloud data in forest sample plots with higher density trees or undergrowth shrubs, it is more likely that trees will block one another and undergrowth shrubs will block the trunks, producing incomplete point cloud data at a certain height which leads to missing or misidentification of trees. Several studies [22,40,42] have shown that the method of slicing point clouds can effectively improve accuracy of octree identification. Unlike existing studies using hierarchical rasterization of collected point clouds, this research directly deals with point cloud data to ensure the accuracy of point clouds and make full use of all acquired data. The thickness of each layer of point cloud is also an important factor that affects tree identification and DBH estimation. In order to ensure the accuracy of DBH estimation, tree diameters were calculated at 1.3 m using multi-layer thickness of point clouds, and the accuracy results are shown in Table 3.

Table 3. Point Cloud Thickness and Accuracy.

Thickness (cm)	RMSE [1]	Number of Trees Detected Correctly	Number of Trees Undetected	Error Detection [2]
1.00	2.92	53	27	22
2.00	3.04	63	17	20
3.00	2.58	75	5	29
4.00	2.99	74	6	14
5.00	2.57	74	6	8
6.00	2.33	75	5	5
7.00	2.53	75	5	7
8.00	2.62	78	2	17
9.00	2.65	78	2	15
10.00	2.74	76	4	12

[1] RMSE: Root mean square error compared with the measurement result; [2] Error Detection: Number of misidentified trees compared with manual recognition.

It can be seen from Table 3 that, when the thickness of slicing point cloud is 6 cm, the RMSE of individual tree DBH is the smallest (2.33 cm). Therefore, point clouds at 0.97 m–2.03 m are sliced into 11 layers with an interval of 0.10 m and a thickness of 0.06 m (0.97 m–1.03 m, 1.07 m–1.13 m, 1.17 m–1.23 m, 1.27 m–1.33 m, 1.37 m–1.43 m, 1.47 m–1.53 m, 1.57 m–1.63 m, 1.67 m–1.73 m, 1.77 m–1.83 m, 1.87 m–1.93 m and 1.97 m–2.03 m).

2.2.3. Octree Segmentation and Connected Component Labeling

In order to reduce redundancy and improve processing efficiency and accuracy, octree segmentation and connected component labeling are combined to segment the point clouds before trunks are identified.

The method of connected component labeling [54] is usually used to detect connected areas of binary images in the field of computer vision. It can be used for processing color images and higher dimensional data as well. Different from the image data, point cloud data is composed of a large number of independent, discrete points with spatial coordinates. Therefore, the method of octree segmentation is used to obtain voxelization data of the hierarchical point cloud. Voxelization is a processing of point cloud segmentation based on octree. First, a closed minimal cube is determined as a root node or a zero-level node, and then the root node is subdivided into eight voxels recursively. Non-empty voxels continue to be divided until they are divided into the remaining thresholds or the minimum pixel size criteria are reached [55].

As shown in Figure 5, the raw point cloud contains a large number of useless points (shrubs, weeds, etc.). With the increasing depth of octree (Figure 5b–h), the points are divided into relatively independent spaces. When the octree level = 10, trunks, shrubs and weeds show better separability. By further increasing the depth of the octree (Octree level = 11 or Octree level = 12), the original separability between the trees is maintained, but the amount of data has increased substantially. Therefore, this study uses the octree segmentation method with octree level = 10 to voxelize each layer of cloud data of trunks. Based on voxelization of points, we use the method of connected component labeling to get point cloud voxels connected and complete the segmentation of tree stem form stratified point clouds. The segmentation results are shown in Figure 5i.

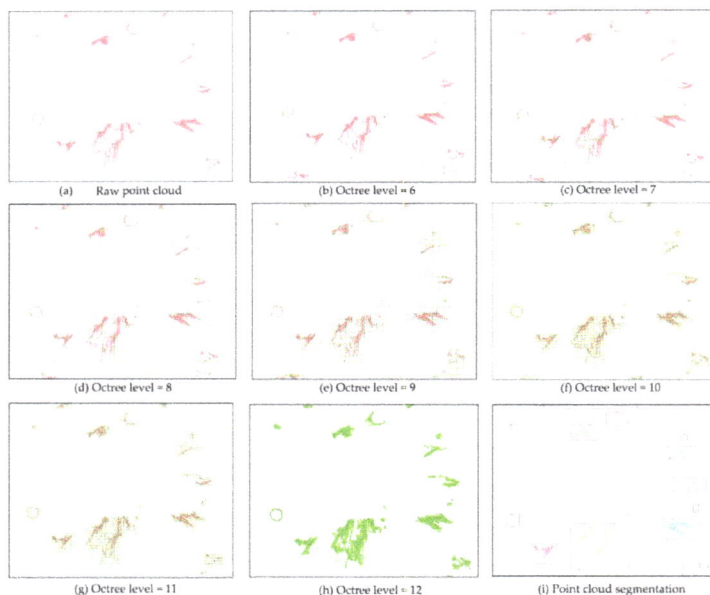

Figure 5. Processing of octree segmentation and connected component labeling (top view). (**a**) Raw point cloud with shrubs and weeds; (**b–h**) With the increasing depth of octree, the points are divided into independent spaces relatively; (**i**) The point cloud is divided into different parts (represented by different colors), and randomly taking points within a single area can effectively reduce the invalid loop.

2.2.4. Random Hough Transform and DBH Extraction

The detailed process of extracting DBH using random Hough transform is shown in Figure 6. For sliced point cloud data, the RHT method is used to sequentially perform circular detection from multiple sub-regions in each layer respectively until extraction of all sliced point cloud is completed. The process of extracting each sub-partition (point cloud set P) of each layer is as follows:

(1) First, the point set P is projected onto the X–Y plane in the direction of Z-axis to form a 2-D point cloud set P' (Figure 6f). Defining the Hough space M (m, n, r) is carried out, where m is the number of grids with 0.01 m intervals for point cloud set P' in the direction of X-axis, n is the number of grids with 0.01 m intervals for P' in the direction of Y-axis, and r is the radius stored in millimeters (Figure 6g, the gray grid under points). Three points $p_1(x_1, y_1)$, $p_2(x_2, y_2)$, $p_3(x_3, y_3)$ that are non-collinear and where the distance between any two points is greater than 0.02 m are selected from the point cloud set P' randomly. The condition of three non-collinear points $p_1(x_1, y_1)$, $p_2(x_2, y_2)$, $p_3(x_3, y_3)$ can be expressed as:

$$\begin{vmatrix} x - x_1 & y - y_1 & z - z_1 \\ x_2 - x_1 & y_2 - y_1 & z_2 - z_1 \\ x_3 - x_1 & y_3 - y_1 & z_3 - z_1 \end{vmatrix} = 0 \tag{1}$$

The distance conditions between the points are:

$$\begin{aligned} \sqrt{(x_1 - x_2)^2 + (y_1 - y_2)^2} &> 0.02 \\ \sqrt{(x_1 - x_3)^2 + (y_1 - y_3)^2} &> 0.02 \\ \sqrt{(x_2 - x_3)^2 + (y_2 - y_3)^2} &> 0.02 \end{aligned} \tag{2}$$

Then, these 3 points can form a circle C_1, with the center point O_1 (a_1, b_1) (Figure 6g) and the radius r_1 of the circle can be obtained. According to our field survey results, if $r_1 > 0.7$ or $r_1 < 0.03$ (trees with DBH larger than 1.40 m or less than 0.06 m are not extracted), a new set of three points should be selected for calculating the radius r_i until r_i satisfies $0.03 \leq r_i \leq 0.7$. The corresponding Hough parameter space is voted in as $M(a_i, b_i, r_i) = M(a_i, b_i, r_i) + 1$.

(2) This method is repeatedly performed on the remaining point clouds until the elements in P' are depleted, so that the final M is obtained. If the difference between the radii of two concentric circles in M is less than 0.01 m, the circles are considered to be the same circle, the average radius of all concentric circles is used as the final radius, and the final voting result is the sum of all circles that meet the conditions. Formula (3) expresses the voting result in M:

$$\frac{M(a_i, b_i, r_i)}{\max(M)} > \varepsilon \tag{3}$$

where, ε is the threshold value of a circle detected for sliced point cloud of trees. Many tests in the study show that the accuracy of DBH extraction is high when $\varepsilon = 0.80$. The next condition needing to be tested is the relative position between any point (x_i, y_i) in point cloud P' and the circle C_i (a_i, b_i, r_i) satisfying the voting result in M:

$$\sqrt{(x_i - a_i)^2 + (y_i - b_i)^2} < 0.7 \times r_i \tag{4}$$

Equation (4) indicates that there are points inside the identified trunk, which are inconsistent with the actual results and should be excluded from the circle that satisfies the voting result.

(3) Using this method, all layers of point clouds are extracted, and the trunk position and the trunk section radius of each layer of trees are obtained. If the position of tree trunk is detected in four or more layers, it is assumed that there is a tree at this position, and the single-wood position is

the center of the trunk closest to the ground. If a trunk can be accurately identified at a height of 1.30 m, DBH of the tree is diameter of the circle identified (Figure 6k). If the trunk cannot be identified, the linear regression method is used to fit the trunk radius and trunk height to obtain DBH (Figure 6i).

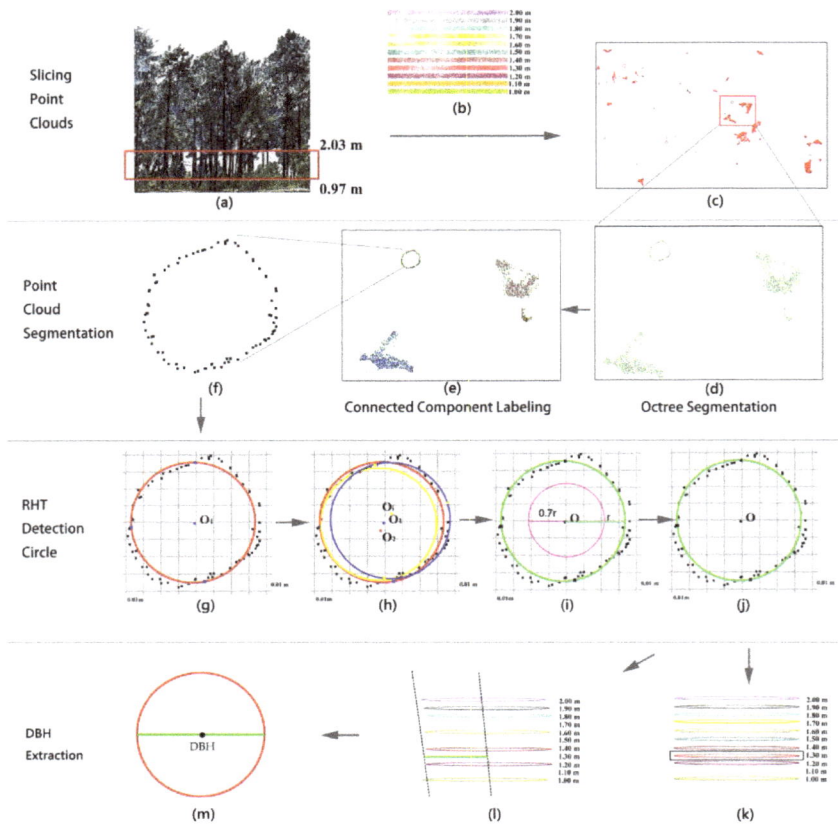

Figure 6. The detailed process of extracting DBH using random Hough transform. After extracting and slicing point cloud data from normalized data, each layer of point cloud is processed separately. Methods of octree segmentation and connected component labeling are used to segment each layer's points. Finally, RHT method is used to extract the trunk and obtain the DBH.

2.2.5. Tree Height Extraction

According to the field surveys, trees in most sample plots in the study area grow in a vertical direction and the trunks are relatively straight. Using the extracted tree locations, point clouds are sliced at an interval of 1.00 m with a thickness of 0.06 m. The RHT method is used to obtain the diameter of multi-layer trunks and the centers of the circles, and the multi-layer centers of circles are used to fit a straight line in the space, which is the growth direction of the tree. Based on the different DBH of each tree, the point clouds are cropped from the bottom to the top along the fitted line within a certain range. These point clouds are considered to be from the same tree, and the height of the tree is considered to be the height of the highest point of all points. However, for sample plots with higher density of forest trees, this method cannot be applied to the lower trees because there may be point clouds of other trees along the growth direction of the trunk, as shown in Figure 7a. To handle such situations, Liu et al. [22] adopted a method of vertical detection along the growth direction of the trunk to calculate the tree height of the lower tree by counting the changes in the voxel of the point cloud. However, the method can only reflect the change of the number of point clouds in the direction of Z-axis, and cannot accurately stratify the different levels of trees. In order to detect the attributions of the tree point cloud effectively, extracted tree points (Figure 7b) are segmented using the CCL method based on the octree segmentation described previously. The segmentation result is shown in Figure 7c. It can be seen from Figure 7c that the algorithm separates points of the low tree and points of high-level tree accurately, and the height of the low tree can be obtained from the highest *z* value of the segmented tree.

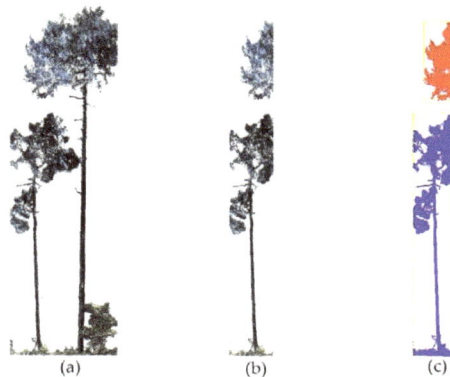

Figure 7. Height extraction of trees in a natural forest. (**a**) Mixture of trees with different heights; (**b**) The height of the highest point of a point cloud may not represent tree height; (**c**) Segmented tree points.

3. Results and Discussion

3.1. Analysis of the Influence of Forest Density on Scanning Range and Accuracy

Forest point clouds collected by TLS are often affected by mutual shelter between trees. Mutual obstruction between trunks results in lower accuracy in tree segmentation and DBH extraction, while mutual shelter between canopies leads to lower accuracy of tree height extraction. From Table 1, it can be seen that the Leica P40 can obtain a large range of high-precision 3D data. However, due to the shelter between trees, the extent of scanning is limited, and the density of trees in forest limits the size of forest sample plots. In order to ensure the accuracy of tree height and DBH extraction, three typical sample plots of *Pinus yunnanensis* (plot numbers 20170726012, 20160831017 and 20160824002) are selected to analyze the accuracy of the same tree species with different forest density (Table 4). According to the result of that, we can determine a range of sample plots suitable.

It can be seen from Figure 8 and Table 4 that topography and forest density affect the scanning range and scanning accuracy of the terrestrial laser scanner.

(1) The scanning range of high-density young forest sample plots is seriously affected by the mutual obstruction between trees. Trees can be identified more accurately (99/106) within a range of 15 m centered on the central station, but there are a small number of missing trees (7 trees) due to mutual shelter between trees within the forest sample plot (5 m–10 m). The identification accuracy of trees near the edge of the young sample plot (distance from the center of the sample plot > 15 m) is low, and there are a large number of missed trees (35). The DBH and extraction accuracy of tree height of the entire sample plot is relatively high (mean RMSE of DBH is 1.03 cm,

and mean RMSE of the tree height is 0.51 m). The maximum error is also located near the edge of the sample plot.

(2) The scanning range of the medium–density plot is mainly affected by the topography and low bushes under the forest canopy. In areas with low tree density and relatively flat terrain, a larger range of scanning areas can be obtained and the accuracy of tree identification and height/DBH extraction are higher as well. For the sample plot of NO. 20160831017, within the range of 20 m from the center of the sample plot, 64 out of 66 trees are identified, with an RMSE of 1.28 cm for DBH, and an RMSE of 0.57 m for tree height. When the distance from the tree to the center of the sample plot exceeds 20 m, the tree recognition accuracy decreases slightly. The tree height and DBH extraction accuracy also slightly decreases with the increase of the distance from the tree to the center of the sample plot.

(3) Low–density mature forests have a relatively complete vertical structure of individual trees. The growth space under the forest canopy is sufficient for the growth of low shrubs. It can be seen from the point clouds (Plot 20160824002) that a large number of shrub points are included in the point cloud near the ground. Meanwhile, the effective range of sample plots obtained by multi–station scanning is limited due to terrain influences. It can be seen from Table 4 that extraction results obtained within the range of 20 m is better than those beyond the range: The tree detection rate is high (36/40), with an RMSE of 1.24 cm for DBH, and an RMSE of 0.46 m for tree height. When the distance from the tree to the TLS scanner is more than 20 m, the accuracy of tree detection is slightly reduced (40/51) due to the longer distance and the influence of shrubs around the station.

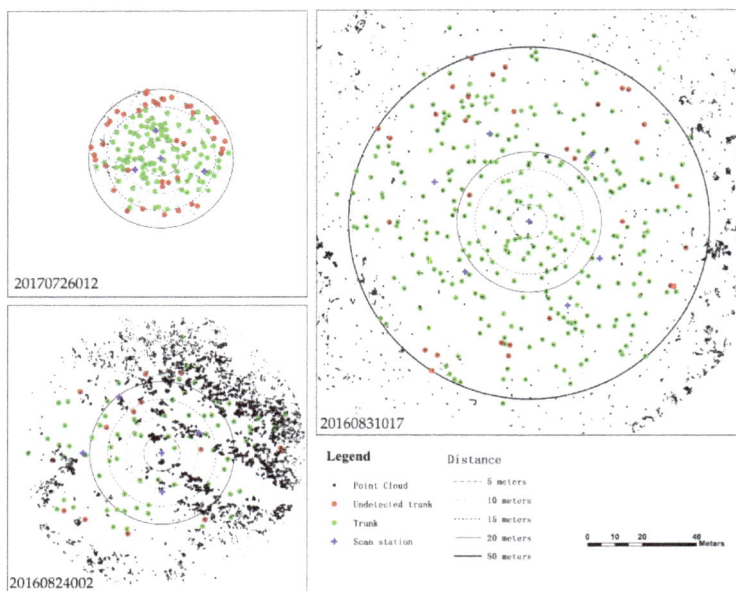

Figure 8. Results of trunk extraction in different density plots of *Pinus yunnanensis*.

Table 4. Effects of Forest Density and Distance on Scanning Range and Accuracy.

Plot #	Stand Age	Mean DBH	Mean T.H.[1]	<5 m				5 m–10 m			
				RMSE		Trees Num.	ER Trees[2]	RMSE		Trees Num.	ER Trees
				DBH	T.H.			DBH	T.H.		
20170726012	Young	11.30	8.2	0.91	0.41	13	0	1.09	0.44	43	7
20160831017	Middle-age	24.70	15.2	1.27	0.54	2	0	1.20	0.89	18	0
20160824002	Mature	28.40	18.0	0.68	0.59	2	0	1.64	0.46	6	0

10 m–15 m				15–20 m				>20 m			
RMSE		Trees Num	ER Trees	RMSE		Trees Num.	ER Trees	RMSE		Trees Num.	ER Trees
DBH	T.H.			DBH	T.H.			DBH	T.H.		
0.81	0.57	43	0	1.30	0.60	12	35	–	–	□	–
1.28	0.56	18	1	1.36	0.30	26	1	1.33	0.78	218	30
1.64	0.52	10	2	1.00	0.17	18	2	1.43	0.84	40	11

[1] T.H.: Tree Height; [2] ER Trees: Error trees, misidentified trees.

3.2. Analysis of the Influence of Forest Types on the Accuracy of Results

The morphological characteristics of forest trees often change with forest species, forest age and growing environment. *Pinus yunnanensis* and *Pinus densata* are genus *Pinus*, and their trunks are mostly straight–lined and their crowns clustered (Figure 9a,b). The trunk of *Picea* and *Abies fabri* grows upright and the lateral branches grow into the surrounding layers. The shape of the tree crown shows an approximate cone (Figure 9c). *Quercus semecarpifolia* is significantly different in morphology from the other three tree species, with a more curved trunk and wider coverage of the crown (Figure 9d). The young, middle-aged and mature forest sample plots of the four types of dominant forest species are selected to analyze the differences in extraction accuracy of tree height and DBH for different forest types and different tree shapes.

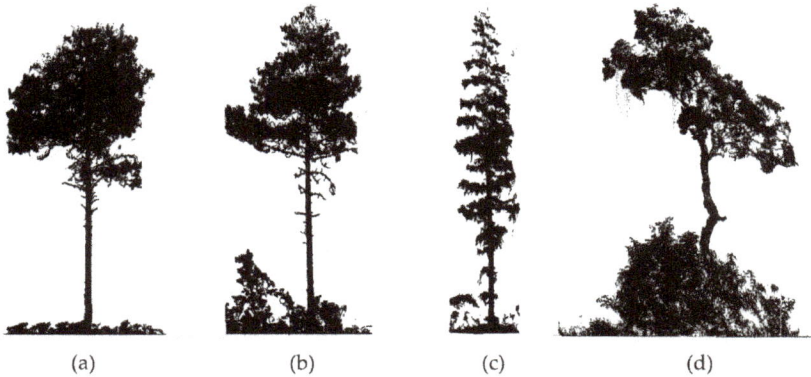

Figure 9. Morphological characteristics of dominant forest tree species in Shangri-La. (**a**) *Pinus yunanensis*; (**b**) *Pinus densata*; (**c**) *Picea* & *Abies fabri*; (**d**) *Quercus semecarpifolia*.

(1)   The influence of forest types on DBH extraction accuracy

It can be seen from Figure 10 that the accuracy of DBH extracted from point cloud data of four main tree species is very high using the method described in Figure 3, with an average RMSE of 1.28 cm, and a minimum error of 1.17 cm for *Pinus yunnanensis*. The RMSE of DBH extracted is 1.52 cm, but the correlation is highest among all forest types ($R^2 = 0.986$). It can be seen from Figure 9 that the four tree species have different morphologies, but the trunks are all nearly circular and are mostly straight near the ground. The results suggest that it is conducive to use the RHT method to extract DBH.

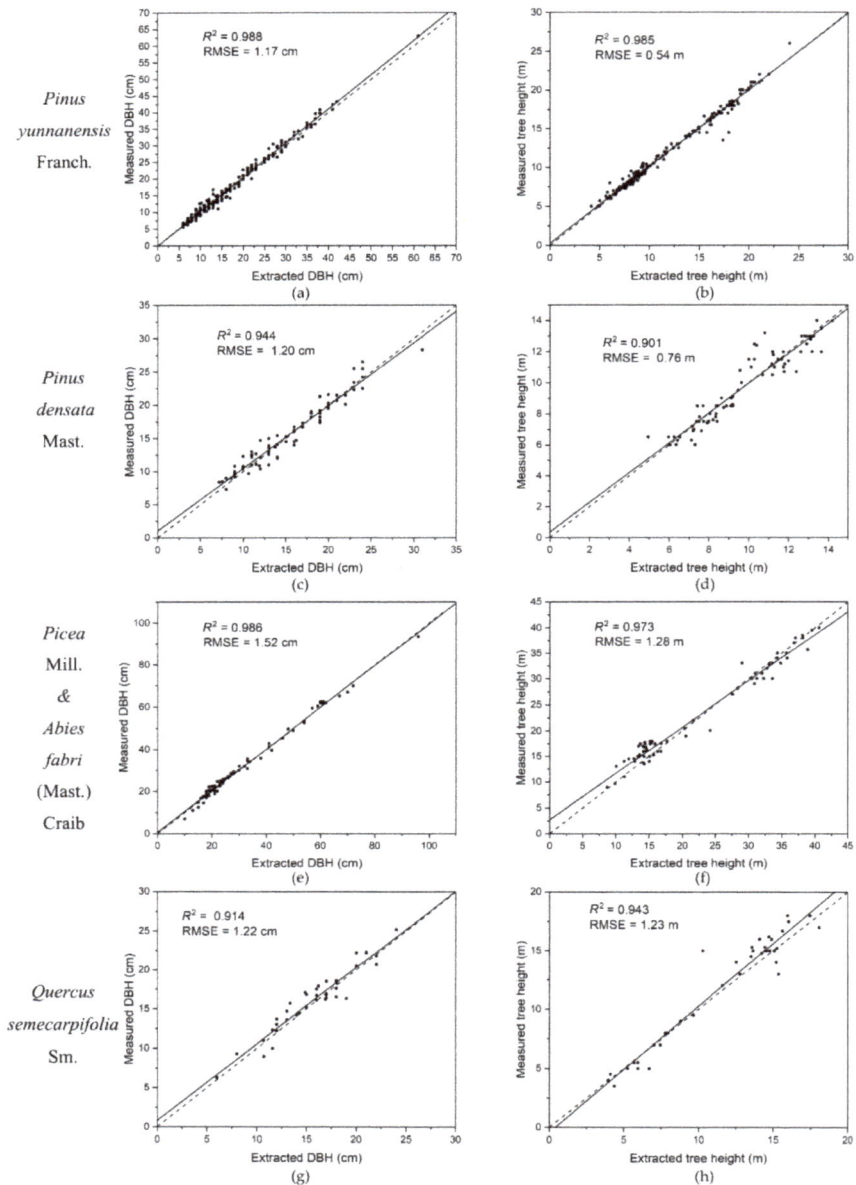

Figure 10. Accuracy analysis of four dominant forest species in Shangri-La. (**a,b**) *Pinus yunanensis*; (**c,d**) *Pinus densata*; (**e,f**) *Picea* & *Abies fabri*; (**g,h**) *Quercus semecarpifolia*.

(2)   The effect of tree species on tree height extraction accuracy

Several studies [38,56,57] have shown that the extraction of forest trees by TLS cannot obtain the point cloud information at the top of the canopy due to the mutual occlusion between canopies, which leads to an underestimation of tree height. The forest in the study area is mainly coniferous with some broad-leaved trees. Compared with broadleaf forests, coniferous forest canopy has some voids, and relatively accurate tree heights can be obtained in a certain range when the stations are properly arranged. It can be seen from Figure 10 that the largest underestimation of tree height is from *Picea & Abies fabri* (RMSE = 1.28 m), followed by *Quercus semecarpifolia*, *Pinus densata* and *Pinus yunanensis*. The mean RMSE of 4 species is 0.95 m, and all of them showed good correlation.

3.3. Accuracy Analysis of Results in Forest Sample Plots

Major elements of forest surveys include tree diameter, height, coverage, and density, among which the tree height and diameter are the most important ones. Mean DBH is the diameter corresponding to the average basal area of dominant tree species, which is a basic index reflecting the forest roughness. The methods of calculating mean DBH include the arithmetic mean method, quadratic mean method, volume mean method, mode method, and median method. At present, the method of quadratic mean is commonly used in forest surveys:

$$\overline{D} = \sqrt{\frac{\sum d_i^2}{n}} \tag{5}$$

where, $\overline{D}$ is the mean DBH, d_i is the DBH of tree i, and n is the total number of trees in forest sample plots.

The mean stand height is an important indicator that reflects the height level of stands, and it is an important tree parameter in forest surveys. For the measurement of arborous forest, it should be determined by selecting 3 to 5 average sample trees among the main forest layer dominant tree species according to the average DBH, and the average tree height should be calculated using the arithmetic mean method.

In this paper, the mean DBH and mean stand height are calculated by using the methods above, and compared with the measured data in sample plots for accuracy assessment. It can be seen from Figure 11 that the mean DBH and mean tree height extracted by the RHT method combined with octree segmentation have strong correlations (correlation of DBH is $R^2 = 0.957$, correlation of tree height is $R^2 = 0.905$) with the measured data. The mean RMSE of the extracting method is 1.96 cm. The smaller the mean DBH, the higher the extraction accuracy. With the increase of the mean DBH in plots, the errors tend to increase slightly. The RMSE of the extracted mean stand height is 1.40 m. Because the canopy obscures the point cloud, results of tree height extracted by TLS data are slightly lower than the actual tree height in forest sample plots with higher average tree heights.

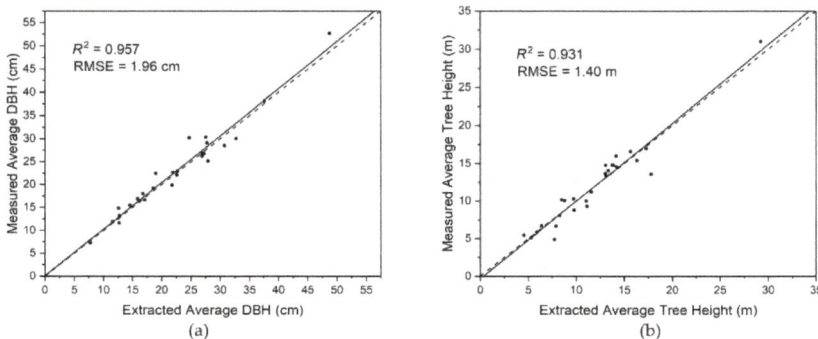

Figure 11. Accuracy analysis of DBH and tree height in forest sample plots.

4. Conclusions

The study focused on extracting tree height and DBH data of natural forest species at plot level in Shangri-La, Northwest Yunnan, China. Combining methods of octree segmentation, CCL and RHT algorithm, tree heights and DBH of natural forests at individual tree level and plot level were obtained. Topography, understory shrubs and tree density influence the TLS scanning range and accuracy of results. Because of different morphology of different tree species in Shangri-La, the accuracy of tree height and DBH extraction for different tree species is different using the method. In general, *Pinus yunnanensis*, *Pinus densata* and *Picea* & *Abies fabria* are coniferous forests, with vertical trunks and similar morphological structures and tree height extraction precision is high. *Quercus semecarpifolia* is a broad–leaved forest species, and its morphology is different from that of coniferous forest, leading to relatively low extraction accuracy. In general, the methods used in the study have high accuracy for the extraction of DBH and tree height for four dominant tree species in Shangri-La. The average RMSE of DBH is 1.28 cm, and the average RMSE of tree height is 0.95 m. The results at forest sample plot levels also show that the method can obtain the mean tree height and mean DBH accurately in complex terrains. In the last few years, mobile/personal laser scanning and image-based techniques have become capable of providing similar 3D point cloud data, and have their own advantages, e.g., lower cost when using image-based techniques and high efficiency when using mobile/personal laser scanning. Further studies need to demonstrate the added value of using TLS, which most probably comes from the highly accurate tree attribute estimates.

Author Contributions: J.W., P.D. and G.L. conceived and designed the experiments; Y.C. performed the experiments; Z.L. analyzed the data; and G.L. wrote the paper. All authors read and approved the final manuscript.

Funding: This research was funded by the National Natural Science Foundation of P.R. China grant number 41271230 and 41561048.

Acknowledgments: Thanks to Cheng Wang of the Institute of Remote Sensing and Digital Earth, Chinese Academy of Sciences for providing guidance in data collection and processing.

Conflicts of Interest: The authors declare no conflict of interest.

References

1. Wu, D.; Li, B.; Yang, A. Estimation of tree height and biomass based on long time series data of landsat. *Eng. Surv. Mapp.* **2017**, 1–5. [CrossRef]
2. Lu, D. Aboveground biomass estimation using landsat TM data in the Brazilian amazon. *Int. J. Remote Sens.* **2005**, *26*, 2509–2525. [CrossRef]
3. Liu, X.; Dan, Z.; Xing, Y. Study on Crown Diameter Extraction and Tree Height Inversion Based on High-resolution Images of UAV. *Cent. South For. Invent. Plan.* **2017**, *36*, 39–43. [CrossRef]
4. Dong, L. New Development of Forest Canopy Height Remote Sensing. *Remote Sens. Technol. Appl.* **2016**, *31*, 833–845. [CrossRef]
5. Ozdemir, I.; Karnieli, A. Predicting forest structural parameters using the image texture derived from worldview-2 multispectral imagery in a dryland forest, Israel. *Int. J. Appl. Earth Obs. Geoinform.* **2011**, *13*, 701–710. [CrossRef]
6. Gibbs, H.K.; Brown, S.; Niles, J.O.; Foley, J.A. Monitoring and estimating tropical forest carbon stocks: Making redd a reality. *Environ. Res. Lett.* **2007**, *2*, 045023. [CrossRef]
7. Chopping, M.; Nolin, A.; Moisen, G.G.; Martonchik, J.V.; Bull, M. Forest canopy height from the multiangle imaging spectroradiometer (MISR) assessed with high resolution discrete return lidar. *Remote Sens. Environ.* **2009**, *113*, 2172–2185. [CrossRef]
8. Rauste, Y. Multi-temporal jers sar data in boreal forest biomass mapping. *Remote Sens. Environ.* **2005**, *97*, 263–275. [CrossRef]
9. Wang, Y. *Estimation of Forest Volume Based on Multi-Source Remote Sensing Data*; Beijing Forestry University: Beijing, China, 2015.

10. Watanabe, M.; Motohka, T.; Thapa, R.B.; Shimada, M. Correlation between L-band SAR Polarimetric Parameters and LiDAR Metrics over a Forested Area. In Proceedings of the 2015 IEEE International Geoscience and Remote Sensing Symposium (IGARSS), Milan, Italy, 26–31 July 2015; pp. 1574–1577. [CrossRef]

11. Solberg, S.; Astrup, R.; Gobakken, T.; Næsset, E.; Weydahl, D.J. Estimating spruce and pine biomass with interferometric X-band SAR. *Remote Sens. Environ.* **2010**, *114*, 2353–2360. [CrossRef]

12. Magnard, C.; Morsdorf, F.; Small, D.; Stilla, U.; Schaepman, M.E.; Meier, E. Single tree identification using airborne multibaseline sar interferometry data. *Remote Sens. Environ.* **2016**, *186*, 567–580. [CrossRef]

13. Wu, Y.; Hong, W.; Wang, Y. The Current Status and Implications of Polarimetric SAR Interferometry. *J. Electron. Inf. Technol.* **2007**, *29*, 1258–1262.

14. Khati, U.; Kumar, S.; Agrawal, S.; Singh, J. Forest height estimation using space-borne polinsar dataset over tropical forests of India. *ESA POLinSAR* **2015**, *4*. [CrossRef]

15. Luo, H.; Chen, E.; Li, Z.; Cao, C. Forest above ground biomass estimation methodology based on polarization coherence tomography. *J. Remote Sens.* **2011**, *15*, 1138–1155. [CrossRef]

16. Schaedel, M.S.; Larson, A.J.; Affleck, D.L.; Belote, R.T.; Goodburn, J.M.; Wright, D.K.; Sutherland, E.K. Long-term precommercial thinning effects on larix occidentalis (western larch) tree and stand characteristics. *Can. J. For. Res.* **2017**, *47*, 861–874. [CrossRef]

17. Huang, K.; Pang, Y.; Shu, Q.; Fu, T. Aboveground forest biomass estimation using ICESat GLAS in Yunnan, China. *J. Remote Sens.* **2013**, *17*, 169–183. [CrossRef]

18. Man, Q.; Dong, P.; Guo, H.; Liu, G.; Shi, R. Light detection and ranging and hyperspectral data for estimation of forest biomass: A review. *J. Appl. Remote Sens.* **2014**, *8*, 081598. [CrossRef]

19. Xing, Y.; Wang, L. ICESat-GLAS Full Waveform-based Study on Forest Canopy Height Retrieval in Sloped Area—A Case Study of Forests in Changbai Mountains, Jilin. *Geomat. Inf. Sci. Wuhan Univ.* **2009**, *34*, 696–700.

20. Nie, S.; Wang, C.; Zeng, H.; Xi, X.; Xia, S. A revised terrain correction method for forest canopy height estimation using icesat/glas data. *ISPRS J. Photogramm. Remote Sens.* **2015**, *108*, 183–190. [CrossRef]

21. Li, Z.; Liu, Q.; Pang, Y. Review on forest parameters inversion using LiDAR. *J. Remote Sens.* **2016**, *20*, 1138–1150. [CrossRef]

22. Liu, L.; Pang, Y.; Li, Z. Individual Tree DBH and Height Estimation Using Terrestrial Laser Scanning (TLS) in A Subtropical Forest. *Sci. Silvae Sin.* **2016**, *52*, 26–37. [CrossRef]

23. Liang, X.; Kankare, V.; Hyyppä, J.; Wang, Y.; Kukko, A.; Haggrén, H.; Yu, X.; Kaartinen, H.; Jaakkola, A.; Guan, F. Terrestrial laser scanning in forest inventories. *ISPRS J. Photogramm. Remote Sens.* **2016**, *115*, 63–77. [CrossRef]

24. Nuttens, T.; De Wulf, A.; Bral, L.; De Wit, B.; Carlier, L.; De Ryck, M.; Stal, C.; Constales, D.; De Backer, H. High Resolution Terrestrial Laser Scanning for Tunnel Deformation Measurements. In Proceedings of the 2010 FIG Congress, Sydney, Australia, 11–16 April 2010.

25. Fröhlich, C.; Mettenleiter, M. Terrestrial laser scanning—New perspectives in 3d surveying. *Int. Arch. Photogramm. Remote Sens. Spat. Inf. Sci.* **2004**, *36*, W2.

26. Telling, J.; Lyda, A.; Hartzell, P.; Glennie, C. Review of earth science research using terrestrial laser scanning. *Earth Sci. Rev.* **2017**, *169*, 35–68. [CrossRef]

27. Buckley, S.J.; Howell, J.; Enge, H.; Kurz, T. Terrestrial laser scanning in geology: Data acquisition, processing and accuracy considerations. *J. Geol. Soc.* **2008**, *165*, 625–638. [CrossRef]

28. Abellán, A.; Jaboyedoff, M.; Oppikofer, T.; Vilaplana, J. Detection of millimetric deformation using a terrestrial laser scanner: Experiment and application to a rockfall event. *Nat. Hazards Earth Syst. Sci.* **2009**, *9*, 365–372. [CrossRef]

29. Prokop, A.; Panholzer, H. Assessing the capability of terrestrial laser scanning for monitoring slow moving landslides. *Nat. Hazards Earth Syst. Sci.* **2009**, *9*, 1921–1928. [CrossRef]

30. Olsen, M.J.; Cheung, K.F.; Yamazaki, Y.; Butcher, S.; Garlock, M.; Yim, S.; McGarity, S.; Robertson, I.; Burgos, L.; Young, Y.L. Damage assessment of the 2010 chile earthquake and tsunami using terrestrial laser scanning. *Earthq. Spectra* **2012**, *28*, S179–S197. [CrossRef]

31. Rosser, N.; Petley, D.; Lim, M.; Dunning, S.; Allison, R. Terrestrial laser scanning for monitoring the process of hard rock coastal cliff erosion. *Q. J. Eng. Geol. Hydrogeol.* **2005**, *38*, 363–375. [CrossRef]

32. Vos, S.; Lindenbergh, R.; de Vries, S.; Aagaard, T.; Deigaard, R.; Fuhrman, D. Coastscan: Continuous monitoring of coastal change using terrestrial laser scanning. In Proceedings of the Coastal Dynamics 2017, Helsingør, Denmark, 12–16 June 2017; Volume 233, pp. 1518–1528.

33. Kuhn, D.; Prüfer, S. Coastal cliff monitoring and analysis of mass wasting processes with the application of terrestrial laser scanning: A case study of Rügen, Germany. *Geomorphology* **2014**, *213*, 153–165. [CrossRef]

34. Anderson, K.E.; Glenn, N.F.; Spaete, L.P.; Shinneman, D.J.; Pilliod, D.S.; Arkle, R.S.; McIlroy, S.K.; Derryberry, D.R. Methodological considerations of terrestrial laser scanning for vegetation monitoring in the sagebrush steppe. *Environ. Monit. Assess.* **2017**, *189*, 578. [CrossRef] [PubMed]

35. Pirotti, F.; Guarnieri, A.; Vettore, A. Ground filtering and vegetation mapping using multi-return terrestrial laser scanning. *ISPRS J. Photogram. Remote Sens.* **2013**, *76*, 56–63. [CrossRef]

36. Vaaja, M.; Hyyppä, J.; Kukko, A.; Kaartinen, H.; Hyyppä, H.; Alho, P. Mapping topography changes and elevation accuracies using a mobile laser scanner. *Remote Sens.* **2011**, *3*, 587–600. [CrossRef]

37. Srinivasan, S.; Popescu, S.C.; Eriksson, M.; Sheridan, R.D.; Ku, N.-W. Terrestrial laser scanning as an effective tool to retrieve tree level height, crown width, and stem diameter. *Remote Sens.* **2015**, *7*, 1877–1896. [CrossRef]

38. Moskal, L.M.; Zheng, G. Retrieving forest inventory variables with terrestrial laser scanning (TLS) in urban heterogeneous forest. *Remote Sens.* **2011**, *4*, 1–20. [CrossRef]

39. Thies, M.; Spiecker, H. Evaluation and future prospects of terrestrial laser scanning for standardized forest inventories. *Forest* **2004**, *2*, 1.

40. Li, D.; Pang, Y.; Yue, C.; Zhao, D.; Xue, G. Extraction of individual tree DBH and height based on terrestrial laser scanner data. *J. Beijing For. Univ.* **2012**, *34*, 79–86. [CrossRef]

41. Bienert, A.; Scheller, S.; Keane, E.; Mullooly, G.; Mohan, F. Application of terrestrial laser scanners for the determination of forest inventory parameters. In Proceedings of the International Archives of Photogrammetry, Remote Sensing and Spatial Information Sciences, Dresden, Germany, 25–27 September 2006; Volume 36.

42. Brolly, G.; Király, G. Algorithms for stem mapping by means of terrestrial laser scanning. *Acta Silvatica et Lignaria Hungarica* **2009**, *5*, 119–130.

43. Shang, R.; Xi, X.; Wang, C.; Wang, X.; Luo, S. Retrieval of individual tree parameters using terrestrial laser scanning data. *Sci. Surv. Mapp.* **2015**, *40*, 78–81. [CrossRef]

44. Janowski, A. The circle object detection with the use of msplit estimation. *E3S Web Conf.* **2018**, *26*, 00014. [CrossRef]

45. Janowski, A.; Bobkowska, K.; Szulwic, J. 3D modelling of cylindrical-shaped objects from lidar data-an assessment based on theoretical modelling and experimental data. *Metrol. Meas. Syst.* **2018**, *25*. [CrossRef]

46. Bobkowska, K.; Szulwic, J.; Tysiąc, P. Bus bays inventory using a terrestrial laser scanning system. *MATEC Web Conf.* **2017**, *122*, 04001. [CrossRef]

47. Cao, T.; Xiao, A.; Wu, L.; Mao, L. Automatic fracture detection based on terrestrial laser scanning data: A new method and case study. *Comput. Geosci.* **2017**, *106*, 209–216. [CrossRef]

48. Wezyk, P.; Koziol, K.; Glista, M.; Pierzchalski, M. Terrestrial laser scanning versus traditional forest inventory first results from the polish forests. *Tanpakushitsu Kakusan Koso Protein Nucleic Acid Enzyme* **2007**, *44*, 325–337.

49. Čerňava, J.; Tuček, J.; Koreň, M.; Mokroš, M. Estimation of diameter at breast height from mobile laser scanning data collected under a heavy forest canopy. *J. For. Sci.* **2017**, *63*, 433–441.

50. Wezyk, P.; Koziol, K.; Glista, M.; Pierzchalski, M. Terrestrial Laser Scanning Versus Traditional Forest Inventory: First Results from the Polish Forests. In Proceedings of the ISPRS Workshop on Laser Scanning, Espoo, Finland, 12–14 September 2007; pp. 12–14.

51. Olofsson, K.; Holmgren, J.; Olsson, H. Tree stem and height measurements using terrestrial laser scanning and the ransac algorithm. *Remote Sens.* **2014**, *6*, 4323–4344. [CrossRef]

52. Zhang, K.; Chen, S.-C.; Whitman, D.; Shyu, M.-L.; Yan, J.; Zhang, C. A progressive morphological filter for removing nonground measurements from airborne lidar data. *IEEE Trans. Geosci. Remote Sens.* **2003**, *41*, 872–882. [CrossRef]

53. Serra, J.; Vincent, L. An overview of morphological filtering. *Circ. Syst. Signal Process.* **1992**, *11*, 47–108. [CrossRef]

54. Dillencourt, M.B.; Samet, H.; Tamminen, M. A general approach to connected-component labeling for arbitrary image representations. *J. ACM* **1992**, *39*, 253–280. [CrossRef]

55. Vo, A.-V.; Truong-Hong, L.; Laefer, D.F.; Bertolotto, M. Octree-based region growing for point cloud segmentation. *ISPRS J. Photogram. Remote Sens.* **2015**, *104*, 88–100. [CrossRef]

56. Király, G.; Brolly, G. Tree height estimation methods for terrestrial laser scanning in a forest reserve. *Int. Arch. Photogram. Remote Sens. Spat. Inf. Sci.* **2007**, *36*, 211–215.

57. Kankare, V.; Holopainen, M.; Vastaranta, M.; Puttonen, E.; Yu, X.; Hyyppä, J.; Vaaja, M.; Hyyppä, H.; Alho, P. Individual tree biomass estimation using terrestrial laser scanning. *ISPRS J. Photogram. Remote Sens.* **2013**, *75*, 64–75. [CrossRef]

MDPI

St. Alban-Anlage 66

4052 Basel

Switzerland

Tel. +41 61 683 77 34

Fax +41 61 302 89 18

www.mdpi.com

Forests Editorial Office

E-mail: forests@mdpi.com

www.mdpi.com/journal/forests